I0474338

SPACE SCIENCE IN THE TWENTY-FIRST CENTURY: IMPERATIVES FOR THE DECADES 1995 TO 2015

MISSION TO PLANET EARTH

Task Group on Earth Sciences
Space Science Board
Commission on Physical Sciences, Mathematics, and Resources
National Research Council

NATIONAL ACADEMY PRESS
Washington, D.C. 1988

National Academy Press   • 2101 Constitution Avenue, N.W.   • Washington. D. C. 20418

Support for this project was provided by Contract NASW 3482 between the National Academy of Sciences and the National Aeronautics and Space Administration.

Library of Congress Catalog Card Number 87-43336

Printed in the United States of America

We now have the technology and the incentive to move boldly forward on a Mission to Planet Earth. We call on the nation to implement an integrated global program using both space-borne and earth-based instrumentation for fundamental research on the origin, evolution, and nature of our planet, its place in our solar system, and its interaction with living things, including mankind.

STEERING GROUP

SPACE SCIENCE BOARD

Foreword

Early in 1984, NASA asked the Space Science Board to undertake a study to determine the principal scientific issues that the disciplines of space science would face during the period from about 1995 to 2015. This request was made partly because NASA expected the Space Station to become available at the beginning of this period, and partly because the missions needed to implement research strategies previously developed by the various committees of the board should have been launched or their development under way by that time. A two-year study was called for. To carry out the study the board put together task groups in earth sciences, planetary and lunar exploration, solar and space physics, astronomy and astrophysics, fundamental physics and chemistry (relativistic gravitation and microgravity sciences), and life sciences. Responsibility for the study was vested in a steering group whose members consisted of the task group chairmen plus other senior representatives of the space science disciplines. To the board's good fortune, distinguished scientists from many countries other than the United States participated in this study.

The findings of the study are published in seven volumes: six task group reports, of which this volume is one, and an overview report of the steering group. I commend this and all the other task group reports to the reader for an understanding of the challenges that confront the space sciences and the insights they promise for the next century. The official recommendations of the study are those to be found in the steering group's overview.

Thomas M. Donahue, Chairman
Space Science Board

Preface

This report outlines a unified program for studying the Earth, from its deep interior to its fluid envelopes. It proposes a system of measuring devices involving both space-based and in situ observations that can accommodate simultaneously a large range of scientific needs.

The scientific objectives served by this integrated infrastructure are cast in a framework of four "grand themes." In summary these are:

1. To determine the composition, structure, dynamics, and evolution of the Earth's crust and deeper interior.

2. To establish and understand the structure, dynamics, and chemistry of the oceans, atmosphere, and cryosphere, and their interaction with the solid Earth.

3. To characterize the history and dynamics of living organisms and their interaction with the environment.

4. To monitor and understand the interaction of human activities with the natural environment.

A focus on these grand themes will help us understand the origin and fate of our planet, and to place it in the context of the solar system.

Chapter 1 introduces the reader to the four grand themes and provides an overview of the Task Group on Earth Sciences' recommended measurement strategy and programs. It emphasizes the need for simultaneous, long-term observations of a wide spectrum of phenomena. The measurements must be global and synoptic, and must consider the time scale of the processes involved. Chapter 2 outlines the present state of knowledge in the framework of the four grand themes. This chapter anticipates the progress that can be achieved by 1995. The scientific objectives for the years 1995 to 2015 are set forth in Chapter 3. The specific sets of measurements needed are integrated with the grand themes to lead to the definition of the required observational capabilities. Chapter 4 reviews some of the observing systems currently operational, being deployed, or planned for 1986 to 1995. Chapter 5 addresses the hardware requirements of the proposed program with emphasis on the satellite Earth Observing System (EOS) and a Permanent Large Array of Terrestrial Observatories (PLATO). The amount of detail varies; in some cases it is expected that specific requirements will be developed by the user community. Chapter 6 concludes with a discussion of science policy considerations and recommendations.

In recent years, a number of important earth science studies have been completed. In particular, the task group has reviewed and utilized several of these reports as the basis for further conclusions relevant to the 1995 to 2015 period. It should be noted at the outset that the task group report is fully supportive of the World Climate Research Program and the International Geosphere-Biosphere Program as described in previous NRC reports. While the task group uses the recommendations of these and other reports, it has gone far beyond them to develop a broader, more comprehensive, and long-range study of Earth for the twenty-first century. The documents referred to in this report are the following:

Data Management and Computation, Volume I: Issues and Recommendations, Committee on Data Management and Computation, Space Science Board, National Academy Press, 1982.
Earth Observing System, NASA Technical Memorandum 86129, August 1984.
Earth Systems Science Committee Working Group on Imaging and Tropospheric Sounding: Final Report, Jet Propulsion Laboratory, Report No. D-2415, January 1985.

Geodynamics in the 1980s, U.S. Geodynamics Committee, Geophysics Research Board, National Academy of Sciences, 1980.

Global Change in the Geosphere-Biosphere: Initial Priorities for an IGBP, U.S. Committee for an International Geosphere-Biosphere Program, National Academy Press, 1986.

The Lithosphere: Report of a Workshop, U.S. Geodynamics Committee, Board on Earth Sciences, National Academy Press, 1983.

Oceanography from Space—A Research Strategy for the Decade 1985-1995, Part 2: Proposed Measurements and Missions, Joint Oceanographic Institutions, Inc., 1984.

A Strategy for Earth Science from Space in the 1980's and 1990's, Part I: Solid Earth and Oceans; Part II: Atmosphere and Interactions with the Solid Earth, Oceans, and Biota, Committee on Earth Sciences, Space Science Board, National Academy Press, 1982 and 1985, respectively.

Finally, I wish to thank the task group for its efforts in preparing the report, and the Space Science Board staff, who provided support for the task group activities.

Don L. Anderson, Chairman
Task Group on Earth Sciences

Contents

1
Design and Implications of a Research Mission to Planet Earth

INTRODUCTION

As we have learned about the other planets of the solar system, it has become more and more evident that Earth, our own planetary home, differs from other planets in several remarkable ways. The first astronauts described Earth as "the blue planet," because of the blue oceans that cover so much of its surface. It might better be called the blue and white planet in recognition of the white clouds that obscure large areas. Our blue and white Earth contrasts sharply with the red of dusty Mars, the dazzling whiteness of Venus, and the complex swirls of pastel colors that characterize Jupiter.

Continued exploration has shown other fundamental differences between planet Earth and all other planets of the solar system. From the human point of view, the most striking of these is that living creatures have existed on Earth for more than 3.5 billion years and have continuously evolved over these eons of time from the simplest one-celled organism to the marvelous diversity of complex life forms that exist today. In contrast, it is almost certain that life is not present today on any of our sister planets and probably never was present during the lifetime of the solar system. Because liquid water is essential for the metabolism and reproduction of living things, the survival and evolution of life on Earth is convincing evidence that our planet has always had a temperature in which water on the surface could remain mostly in liquid form.

Equally fundamental has been the very existence of large quantities of water on the outer surface of Earth, quite unlike her sister planets, Mars, Venus, or Mercury. If the liquid oceans were not present, several other of Earth's unique characteristics that make life and its evolution possible could not exist. That we have relatively modest amounts of carbon dioxide in our atmosphere, and hence have avoided the runaway greenhouse effect that makes Venus uninhabitable, results from the fact that nearly all the carbon dioxide that has flowed out of Earth's interior during her lifetime has been buried in ocean sediments as limestone or as organic carbon produced from atmospheric carbon dioxide by photosynthesis.

The presence of free oxygen would not be possible without the photodissociation of water and the consequent escape of hydrogen. Without the presence of oxygen, ultraviolet-shielding ozone would not exist in the stratosphere and life on land would be impossible. Finally, most animals could not exist, either on land or in the sea, without the energetic metabolism made possible by free oxygen.

Liquid water and carbon dioxide, acting together, transform rocks by weathering into clays, which were perhaps the template for life, and into soluble substances that are among the essential inorganic nutrients for plants. These include phosphorus and potassium, and many trace substances.

As we think more deeply, we realize that these processes must be limited by other, more subtle effects, or else life could not persist. Oxygen in moderate amounts is a necessity for animal life, but in higher concentrations is a poison. If oxygen continued to accumulate in the atmosphere, fires and other kinds of rapid oxidation would destroy all living things. If organic matter continued to accumulate in the deep-sea sediments, all the nutrients released by weathering would eventually return to insoluble forms and Earth's plants would starve. Similarly, if limestone sediments continued to accumulate in the ocean without a compensating inflow of carbon dioxide from the deep-sea ridges and other volcanic eruptions, the concentration of atmospheric carbon dioxide could become so low that photosynthesis would be impossible.

We thus come to one of the most remarkable phenomena

on Earth and in the solar system—plate tectonics—the continual recycling of Earth's surface materials deep into the interior, and their reappearance at mid-ocean ridges and volcanoes. This process of tectonic renewal, apparently unique to Earth, may be essential to the persistence of the benign environment that has allowed life to exist and evolve in all its diversity for over 3.5 billion years. Motions in the deep interior drive the plates and generate the magnetic field that partially shields the Earth from the harsh environment of space.

Astronomical chance forms the framework of this benign environment. If the Earth were much smaller, it could not retain an atmosphere. If it were much closer to or much further from the Sun, the oceans would boil or freeze. If its orbit and axis of rotation did not fluctuate, the cyclical variations in climate that have spurred evolution would not exist. If the Sun were a binary, a subtle orbit of uniform conditions for the Earth would be very unlikely. But it is Earth's own inner life—the convective processes deep within its interior, perhaps largely controlled by the flux of heat from radioactive decay and to an unknown extent by the primordial heat of agglomeration—that has determined its history and our own.

Why does the phenomenon of plate tectonics exist on Earth but not on Venus, a near twin of Earth? What are the characteristics of Earth that make plate tectonic convection possible? Is it entirely the low surface temperature that makes it possible to recycle water as well as rock? What is the nature of the convective process; how have the rates of convection varied with the general decline in energy of the system; what are the rates of interchange between parts of the mantle? What are the effects of changing rates of convection on the Earth's surface, for example on atmospheric carbon dioxide concentration, and hence on climate and life itself? What insights can we gain from studies of the variable magnetic field generated by Earth's core-mantle dynamo?

Even the question of the origin of life may be related to plate tectonic processes. One of the most remarkable findings of recent years has been that of the existence of complex ecosystems of fishes, invertebrates, and bacteria around the deep-sea vents in the mid-ocean ridges. In these vents, water at temperatures of several hundred degrees centigrade exists in liquid form because of the high hydrostatic pressure. In this hot liquid water there is evidence that there are living anaerobic sulfide-oxidizing bacteria, which provide (together with their relatives living symbiotically within the animals) the energy and organic compounds for the large variety of animal inhabitants of the vent communities. Here is an environment that may well have been the seat of life's origin on Earth, despite its inaccessibility to photosynthesis. High temperatures would have allowed rapid chemical reactions, and reduced sulfur compounds could have provided a rich source of energy and, eventually, the mineral resources upon which mankind is dependent. The great mass of overlying water would have given complete protection from destructive ultraviolet radiation. Even the composition and concentration of the salts in seawater may be determined by plate tectonic processes, specifically the hydrothermal circulation that occurs in a wide zone between the ocean and the upper lithosphere on each side of the mid-ocean ridges.

Another unanswered question is the nature of the Earth's response to the asteroid and comet collisions that strike the Earth at intervals. What has been their effect on the evolution of life? Some have suggested that "great extinctions" due to these collisions have stimulated the rapid evolution of new living forms. The search for evidence of such collisions during the geologic past may throw a new light on evolutionary processes.

Our present global view of the Earth has been synthesized from decades of painfully collected regional and local data, operational weather satellites, and a few tantalizing pilot programs that have mapped some properties of the ocean and land surface from space. This is in contrast to our missions to other planets, which, from the beginning, provided global, integrated, and simultaneous measurements. The Earth is the only planet on which we can simultaneously make global satellite observations and deploy adequate instrumentation to image the interior, including the mantle and core, in order to address such fundamental questions as the origin of the magnetic field and the nature of the convective overturning of the Earth's mantle and crust.

In this report the task group proposes a Mission to Planet Earth as an essential part of our country's program of space exploration. This proposal is not as paradoxical as it sounds, for it concerns the integrity and unity of Earth as a planet, and it emphasizes the necessity for studying Earth as a whole. Only by a mission to Earth can we obtain a satisfactory degree of resolution of the changes in the earth environment over time, and therefore of history. Moreover, only by studying Earth can we begin to understand the relations among several unique phenomena.

The task group therefore proposes a Mission to Planet Earth to include all the program elements required to understand a planet with an atmosphere, hydrosphere, biosphere, solid crust, mantle, and solid-liquid core. The proposal sets forth a concerted and integrated research program on the origin, evolution, and nature of our planet and its place in the solar system.

GRAND THEMES

The primary research objectives are addressed by four "grand themes" that are developed at length in this report:

1. *To determine the composition, structure, and dynamics of the Earth's interior and crust, and to understand the processes by which the Earth evolved to its present state.*

The Earth's crustal surface is the home of man and the interface between the rapid variation in the fluid envelopes and the usually slow, sometimes catastrophic motions of the interior. The crust contains the record of past events on Earth, which is the main object of the study of geology.

The Earth's mantle is undoubtedly in a state of thermal convection, but such important properties as its composition, the spectrum of convective scales, the degree of interaction between upper and lower mantle, and its relationship to volcanism and tectonics are imperfectly known. Considerable improvement in understanding its effects on the crust and lithosphere (e.g., earthquakes, igneous differentiation) are attainable. Loosely coupled to the mantle is the fluid outer core, the source of the time-varying magnetic field. Closer measurement of this variation, with seismic imaging of the interior and computer modeling, should constrain the nature of the geodynamo. This entire system of the Earth's interior is evolving in a general trend of decline in energy and compositional stratification, but the rate of this decline and oscillations about the trend are poorly known. The starting conditions for this evolution depend on the formation of the Earth and the other planets from the solar nebula and thus constitute a significantly different problem.

2. *To establish and understand the structure, dynamics, and chemistry of the oceans, atmosphere, and cryosphere and their*

interactions with the solid Earth, including the global hydrological cycle, weather, and climate.

The factors underlying the Earth's energy budget—temperature, precipitation patterns, sea level rise, and other properties—are not well enough known to predict climate confidently. Measurements of carbon dioxide and other atmospheric gases, dust and aerosols, and related phenomena, coupled with improved modeling, are needed. Improved understanding of oceanic circulation and its effects on climate should be achievable. The land surface needs to be more systematically monitored to map and establish trends in surface composition, tectonics, soil erosion and salinization, geomorphology, vegetation (state, as well as distribution of types), hydrologic phenomena (snow cover, ground water), and other properties. Atmospheric dynamical processes, including air-sea interaction, can be better measured and modeled. The effects of biological processes on the hydrological cycle, climate dynamics, and geochemistry are major problems as discussed below.

3. *To characterize the interactions of living organisms among themselves and with the physical environment, including their effects on the composition, dynamics, and evolution of the ocean, atmosphere, and crust.*

The biosphere is an important part of the fluid outer layers of the Earth, controlling the oxygen content and other factors. Waxing and waning of the biota in the ocean are major factors in its changes; this ocean ecosystem, as well as that of land, needs to be better understood, both globally and locally. Conversely, biological evolution is influenced by the physical environment in a variety of ways, including climate, continental drift, and asteroid impacts. Thus, this theme is closely tied to the one above.

4. *To monitor and understand the interaction of human activities with the natural environment.*

The impacts of population increase, agricultural and industrial development, and energy consumption on the natural environment are subjects of great scientific interest, as well as practical concern. Human activity is clearly affecting gases such as carbon dioxide and methane as well as the dust content of the atmosphere. Population increases in developing countries are contributing to the rate of desertification and urbanization. Tropical deforestation has important implications as to climate and genetic diversity. Industrial effluents are blighting temperate zone forests. Conversely, many developments have made mankind more vulnerable to natural hazards such as storms and earthquakes. Most of these trends are best monitored from space.

Considering the themes as a whole, the task group notes that the complexity of the Earth on a global scale makes any division, such as these grand themes, to some extent arbitrary. The atmosphere has a time scale of hours to days for evolution of weather systems, but its climatic conditions are obviously buffered by the ocean, which has inherent time scales of months to centuries. Meanwhile, both systems are greatly influenced by the biosphere, which can undergo changes measurable from space on time scales ranging from days to decades and longer. The influences of the Sun have an established 11-year cycle, but underlying it are variations and trends on century and millennium time scales. The cryosphere underwent a great decline 8,000 years ago, but the geologic record suggests a 100,000-year time scale for its waxings and wanings. There appear also to be oscillations on the scale of a few hundred years (e.g., the "little ice age" of the seventeenth century) and some thousands of years, the latter associated with orbital dynamics. The interaction of the mantle with the crust and lithosphere is believed to undergo changes in character on a time scale of 10 million years. In tectonically active zones a given fault may have an earthquake every 200 ± 100 years. A volcano may outburst on a similar interval, sometimes with global climatic consequences. This great welter of causative effects with different time scales requires measurements by a variety of means, all requiring completeness, simultaneity, and continuity.

MEASUREMENT STRATEGY

The issues of overall measurement strategy for an earth observing system have been considered in detail in a number of previous reports, including the two volumes of *A Strategy for Earth Science from Space in the 1980's and 1990's* by the Committee on Earth Sciences (CES) of the Space Science Board (1982, 1985) and the report *Oceanography from Space—A Research Strategy for the Decade 1985-1995* by Joint Oceanographic Institutions (1984). These reports note, and this task group agrees, that three basic themes are fundamental to advances in our understanding of the causes and effects of global change. The measurements must be *global and synoptic*, they must be carried out over the *long term*, and different processes such as solar output, winds, currents, and geological and biological activity must be measured *simultaneously* (to a degree dependent on the rates involved). The first theme is the *global and synoptic* nature of measurements. We have learned that advances in earth science derive from the synthesis of new ideas that come from global synoptic observations. For example, we owe our new understanding of plate tectonics or large-scale atmospheric circulation to global observations and models. In each case, the global observations are a synoptic view, that is, a snapshot that gives a picture of the system over a period that is short compared to the time over which the system changes. For example, the relevant time scales of the ocean— months to decades—fall between geologic periods and atmospheric weather events.

The second theme of the measurement strategy is *long-term continuity*. The Earth as a system is energetic on many scales, from microseisms to interannual El Niños to ice ages. However, even if we restrict ourselves to decadal time changes—the human time scale—statistics dictate that measurements over many years will be required before we can make accurate statements about the energetics of the system.

As we look into the future, we see the need for geodynamic, climate, and biosphere measurements for decades, and we can see that the addition of gravity and magnetic fields, continental drift, and solar-terrestrial interactions to these processes extends the necessary observational time periods to centuries. However, to understand the system we need more than long-term measurements. We must at the same time make clear the fundamental processes at work. The results from these process studies will be used to help put the whole picture together. Unfortunately, the time periods for the required observations are much longer than the time scales for political decision-making. Thus, there is a need for a national commitment to carrying measurements through the necessary time periods.

The third theme is *simultaneity*—that is, observations of different kinds of processes at the same time. This includes study of the land, ocean, atmosphere, and biota as an interactive system on a global scale. Earth sciences have tended to treat these components as separate disciplines. Much is known about, for example, the processes that govern winds and temperature in the atmosphere, geological processes, or the chemistry of trace substances in the ocean. In the last few decades, however, increasing attention has become focused on questions that transcend traditional disciplinary boundaries and require, in addition, an understanding of the complex linkages and feedbacks between these components. As is noted in the CES reports, the changes to be expected from the worldwide deforestation and consumption of fossil fuels, from increased erosion of continents, from the sensitivity of the stratospheric ozone to trace gases such as chlorofluorocarbons, and from the causes of past and present extinctions of whole classes of living species are all questions of fundamental importance. They can only be addressed from a global, interdisciplinary perspective, drawing on a wide spectrum of observations and skills beyond the range of any one individual, institution, or agency.

Observing and understanding such a complex system is a basic intellectual challenge. Yet these questions and others like them are also issues that have to be resolved if we are to predict the future, or even diagnose correctly the changes that are under way around us. Some of them are results of activities of humankind, others are natural fluctuations for which precedents are undocumented or simply unrecognized. The planet Earth is our environment and, for better or for worse, we are part of that environment, reacting to it and acting upon it in ways that are far-reaching but as yet barely perceived. It behooves us to pay attention.

Obtaining such an understanding requires long-term study. In particular, the interactions between the parts are crucial. Boundary layers and boundary phenomena are of special interest, since this is where such interactions take place. This implies building an information base including sustained global observations, and evolving quantitative models to be used for examining responses and feedbacks as well as predicting the behavior of the whole. The required information base is large and diverse. Some aspects, such as long records of surface temperature at land stations in northern temperate latitudes, already exist, though not necessarily in a form that is readily accessible; for others, modifications of existing data sources are appropriate. However, for some aspects, particularly those providing global coverage, the deployment of new observational systems is required. The coverage and uniform data quality obtainable imply a critical role for satellite-based remote sensing, but extensive in situ measurements are also needed, distributed in key locations around the globe.

SYSTEMS OVERVIEW

The task group's specific recommendations are for implementation of a system for observing the Earth that builds on and expands the Earth Observing System (EOS) that is currently planned to fly as part of the Space Station complex in the mid-1990s. EOS in turn will build on predecessor missions, both domestic and foreign, which will help define the specific parameters and orbits needed for adequate long-term monitoring of Earth. EOS will be the first phase in the development of long-term satellite measurement systems. Here the task group looks beyond the initial deployment of EOS to lay out a series of specific recommendations as to the overall structure and programmatic content of a long-term global mapping and monitoring system for Earth, including satellite and in situ systems.

1. A Satellite-based Observing System

The task group recommends that the centerpiece of the global observing system be a network of satellites and platforms in the following arrangement:

• *A set of five geostationary satellites*, designed to carry a wide variety of instruments to cover the entire Earth for long-term measurements. Five are required to cover the Earth completely to 60° latitude north and south. These satellites would be large, high-power spacecraft, designed to provide continuous measurements of every part of the Earth visible from geostationary orbit. They would carry improved versions of instruments currently used or being developed, plus new technology that would expand our view in space, time, and the energy spectrum (e.g., side-looking imagery) and would be used together with:

• *A set of two to six polar-orbiting platforms* to cover the polar areas, above latitude 60°, and to provide platforms for a variety of instruments that must be closer to the Earth. These polar-orbiting platforms would operate continuously at altitudes of about 824 km, being replaced as necessary, and would carry a wide spectrum of instruments together with common power supplies, data handling, and communications. The number of polar-orbiting platforms is based on a compromise between temporal and spatial coverage and cost; two is the minimum to achieve biweekly coverage of the processes believed significant to cause global change (e.g., 200-km ocean eddies); six would be required for daily coverage. The data from the instruments on the polar platforms and the geostationary satellites would thus yield the required global synoptic data, which when appropriately processed would provide the fundamental long-term data set required for monitoring global change. However, these two sets of spacecraft must be augmented by:

• *A series of special missions*, which require other orbits. Some would be short-term duration (4 days (Shuttle) to 1 to 3 years (Explorer-type missions)); others would be essentially permanent, such as a second-generation Global Positioning System (GPS) constellation. Some of the short-term missions would test instruments and concepts for incorporation into the long-term satellite network discussed above; others would undertake one-time mappings such as improved geological spectrometry and high-resolution terrain mapping.

2. Complementary In Situ Observing Systems

The task group recommends the development, deployment, and long-term operation of a system of in situ measuring devices—the Permanent Large Array of Terrestrial Observatories (PLATO)— to provide complementary data to the space network. Wherever applicable, the data should be transmitted in real time and integrated with observations from space.

In situ measurements represent an essential element of any observing system designed to investigate the Earth as a planet. There is a need to measure effects that cannot be detected through remote sensing from space, to provide increased resolution in regional studies, and to supply calibration and verification of space observations. In situ measurements from PLATO would include, for example, detailed studies of terrestrial and oceanic biomes, ocean bottom stations designed to monitor pressure, seismic and acoustic signals, a variety of probes ranging from balloons to boreholes, and GPS receivers and laser corner reflectors for monitoring tectonic deformation. The number of sites for these instruments would range from 100s to 10,000s; their distribution typically would be nonuniform in accordance with the problem under study.

3. Modeling

The task group recommends that state-of-the-art computing technology be utilized for data analysis and theoretical modeling of earth processes. We are dealing with a complex, turbulent system of living and nonliving material on many scales, involving a wide variety of substances and properties, and ranging from the core to the outer atmosphere. Modeling the system requires the best data sets possible, the fastest computers, and imaginative ideas from researchers. In turn, this modeling can give a context and direction for future observations. The task group sees modeling as an integral part of this enterprise, and the need for the state-of-the-art technology in modeling as critical.

4. Data System

The task group recommends that a full and coordinated data system, which both archives and disseminates data, be established. The task group does not see the necessity for a central archiving of all data, but does see the need for a central data authority to establish formats and other conventions, to identify data location, and to arrange for access to all data as required. We can expect that data rates will be very high, on the order of 10^{13} to 10^{14} bits per day. This will require much selective averaging and heavy use of new technology, possibly beyond video disks to new storage and retrieval technologies. Automation of some phases of the selection and averaging process will be necessary. In short, state-of-the-art technology should be made available for all phases of data handling. A more detailed description of the full Mission to Planet Earth system is provided in Chapter 5.

STRATEGY OVERVIEW

To be effective, an earth research system must be conceived and evaluated as an integrated whole, including contributions from other nations. Consequently, a major effort to achieve these objectives is timely. As world population and economic aspirations press harder on finite resources, the need for improved understanding grows year by year, and environmental problems are transformed into ever more prominent social and political problems, while becoming less amenable to successful solutions.

During the next decade a number of new programmatic initiatives will pace the progress in earth sciences. The World Climate Research Program (WCRP), currently in operation, addresses the physical basis for regional and global climate variability from a few weeks to several years. For the study of the solid Earth, the International Lithosphere Project (ILP) is a coordinated, ongoing international activity to improve our understanding of the structure and dynamics of the outer layers of the Earth. In addition, the Ocean Drilling Program probes the Earth's crust beneath the sea. Plans are being developed for a Global Seismic Network, which will investigate the three-dimensional structure of the crust, mantle, and core. The International Geosphere-Biosphere Program (IGBP), now in the planning stages, will study global change with a focus on the interactive ocean-atmosphere-land-biota system on a time scale of decades to centuries. The success of these ongoing and planned programs is keyed to our ability to obtain accurate global and repetitive data on certain geophysical parameters from satellites, and to complement the space data with reliable surface-based observations, as well as to develop realistic, interactive models that can be tested by diagnostic measurements.

Towards 1995, as the results from these identified programs begin to reach fruition, we will be getting ready to address one of the most fundamental questions in earth sciences; namely, how does the planet Earth as a whole evolve and what path has it followed to reach its present state of evolution? It is clear that in order to succeed we will have to assure that research programs for understanding the evolution of atmosphere, oceans, land, and interior of the Earth are all proceeding in conjunction. We must also assure that satellite and earth-based observing systems put in place for one discipline of earth sciences are coordinated with those developed for the other, so that common observatories for interdisciplinary measurements are the rule rather than the exception.

It is for this reason that the task group is proposing a Mission to Planet Earth that will:

- address the questions that concern the integrated functioning of Earth as a system;
- carry out measurements from space in concert with data from ground-based techniques;

14

- be capable of observing the composition of the atmosphere; the composition, structure, and texture of land surfaces; the biology of the land and oceans; the distribution of land and sea ice; and the structure and dynamics of the interior of the Earth;
- provide long-term, consistent data records of the dynamics of climate, lithospheric plates, biosphere, and the oceans over periods of decades or more;
- be compatible with our needs for monitoring certain phenomena with long time scales (e.g., ocean temperatures), and performing repeat measurements over certain areas for assessing changes because of catastrophic phenomena (e.g., earthquakes), and at the same time be flexible enough to be able to observe targets of opportunity (e.g., a sudden volcanic explosion);
- assure the compatibility and continuity of the current operational and research satellite observing systems;
- develop new instruments in response to measurement needs for parameters that are not currently measurable from space or from the ground; and
- guarantee the uniformity of the data acquisition and archiving system so as to facilitate their use by the research community, and ensure that future improvements in data processing technology can be applied with a minimum impact on the continuity of the data streams.

From this description it should be evident that the solution of the major problems in earth science requires an integrated approach. The key to progress on interdisciplinary issues in earth science during the decade of the 1990s and beyond will be addressing questions that concern the functioning of the Earth as a system.

The overall strategy therefore contains important prerequisites for U.S. science policy from both domestic and international perspectives. These are:

- *A long-term commitment to a vigorous, systematic exploration of the Earth.*
- *Maintenance of an open-skies policy.*
- *International cooperation and coordination regarding all relevant systems, beginning with the early stages of program planning.*
- *Full coordination of the U.S. federal agencies involved in the civil earth science effort at the programmatic and budgetary levels.*

- *Development and support of a more comprehensive program in the solid earth sciences within NASA.*

These science policy issues are addressed in more detail in Chapter 6. Finally, the strategies of the Space Science Board's Committee on Earth Sciences (CES) have emphasized that global earth science investigations from space will naturally be followed by global exploration for resources and by development of natural-hazard warning systems. Of course, all such applications must be based on thorough scientific understanding. The task group hopes that the scientific return derived from the strategy contained in the CES reports and in this report will provide a sufficient foundation for all applications.

2
Earth Sciences—
Status of Understanding

INTRODUCTION

The Earth is more fascinating and mysterious than ever, despite the great advances in knowledge achieved in the first three decades of the space age. The fascination is generated by the extraordinary complexity of the Earth and by the inherent inaccessibility of many of its key processes. In the face of this complexity and inaccessibility, the notion of an orderly progression from reconnaissance to mapping becomes a myth. This is all the more reason to undertake a systematic and comprehensive program. By 1995 we will be ready to make an integrated study of the Earth as a planet, that is, to undertake a Mission to Planet Earth.

Several developments, both recent and expected in the near future, make this timely. The Earth's complexity involves regimes of widely differing energy and time scales interacting in varied ways. For this reason problems of earth science cannot be reduced to fundamental elements analogous to the energetic particles of modern physics or the DNA components of modern biology. Because the systems are "chaotic," it is usually impossible to predict the behavior of a regime within the Earth solely from first principles; attempts to do so are generally less successful than alternative approaches, which may strike the basic scientist as crudely empirical.

The geologic record of a long series of events that actually occurred can be used to forecast Earth's behavior. It is true that some of the interfaces between greatly differing regimes of the Earth are quite sharp, such as the ocean and atmosphere, the ocean and crust, or an organism and its environment. But one of the greatest weaknesses in our understanding concerns what happens at these interfaces. Across these interfaces chemical and energetic fluxes influence behavior on time scales ranging from seconds to millions of years.

Earth is the only planet in our solar system on which life has come into existence and persisted. Why? Not only does Earth support life, it is influenced by life. Biological processes affect the Earth's atmosphere, oceans, and solid surfaces; systematic phenomena such as climate and the global cycling of chemicals respond to life. Conversely, living organisms are influenced directly by the climate system, by the distribution and flux of chemical compounds. As such, the earth system is strongly coupled with widely varying rate constants. The Earth is the only planet that supports plate tectonics. Why? The Earth is the only planet with a liquid ocean. Again, why?

The complexity of the Earth has led to its examination being divided among several disciplines that speak imperfectly to each other. They range from geomagnetic theory to ecology, which is concerned with the interaction between organisms and their environments. One matter in common among these disciplines is the problem of inference from incomplete data. In large part, this problem arises from the inaccessibility of key processes. Extreme examples are convective flows in the lower mantle generated by inhomogeneity of density on the one hand, and the secretion of calcium carbonate by foraminifera in the ocean on the other. To help solve these problems of identifying and quantifying the forces behind sketchily sampled details, we need global, synoptic, and continuous data.

The basic objectives of studies of Earth can be grouped as follows:

- To understand the processes by which the Earth formed and evolved to its present state, and to determine the composition, structure, and dynamics of the solid planet.
- To establish and understand the structure and dynamics of the oceans and atmosphere and their interactions with the

solid Earth including the global hydrological cycle, weather, and climate.

• To characterize the history and dynamics of living organisms, including mankind and their interactions with the environment.

• To understand Earth in the context of the solar system, and the use of the Earth as a detector of cosmic events.

There has to date been no systematic attack on these broad objectives. In this document the task group hopes to show how these objectives can be developed into "grand themes" to focus a systematic, global study—a Mission to Planet Earth. To set the context, the following sections summarize the state of understanding of the various subsystems of the Earth as the task group expects it to be in 1995. In many cases, the task group has recognized the existence of up-to-date reviews and recommendations in previous reports, and has quoted summaries of these in appropriate contexts.

THE EARTH'S INTERIOR AND CRUST

Our primary objective is to understand the processes by which the Earth formed and evolved to its present state, and to determine the composition, structure, and dynamics of the solid planet. Since the synthesis of plate tectonics has given us a new understanding of Earth processes, the discussion will begin there.

Plate Tectonics

As has been pointed out in Part I of the CES strategy, geology has been revolutionized since the mid-1960s by the recognition of the plate structure of the lithosphere. According to the plate tectonic theory, the Earth's surface is divided into about 11 major and a large number of minor plates that behave as rigid units, are in continuous relative motion, and interact mainly at their edges. New plate material is created at ocean ridges; old oceanic plate material is subducted or consumed at ocean trenches. Many active volcanoes are associated with plate boundaries. Earthquakes occur where plates are created or destroyed, and where plates move past one another. Earthquakes outline the world's major plates and serve as energetic sources to probe the interior.

The relative motion of the Earth's plates over approximately one hundred 100,000-year intervals is known for the last 200 million years from studies of magnetic lineations on the ocean floor. These motions give us some idea of the general rates of convective motions in the Earth's viscous mantle. There is no ocean floor older than approximately 200 million years; most older oceanic crustal material has been subducted, some has been incorporated into continents. Inferences about plate interaction prior to this time must be made from continental geology and especially preserved pieces of ocean floor (ophiolite suites).

Although we know the average relative velocities of the Earth's plates over a time scale of a million years, the Earth's magnetic field does not reverse polarity frequently enough to allow a finer resolution of the present rates of motion. We do not know what drives the plates. Earthquakes show that the motion of the plates at plate boundaries is episodic, but we do not yet know how strain accumulates at those boundaries. Nor do we know whether precursory effects before major earthquakes are general phenomena, or diagnostic signals. The episodic motions at plate boundaries are thought to be damped out with distance from the boundary by stress relaxation in the viscous asthenosphere underlying the plates, so that the relative motions of plate interiors are steady; direct observations of plate motions over time scales of years are beginning to indicate that these rates are indeed steady.

A major goal in plate dynamics is to understand the driving mechanism for the plate motions. This mechanism involves some form of thermal convection in the Earth's mantle, but the form of the motions is uncertain at levels deeper than the plates themselves. We do not know whether the radial extent of the convection system involving the plates extends to the full depth of the mantle, or part way. Nor do we know the planform of the flow, that is, the pattern in plan view of upwelling and downwelling limbs of the convection system. Further, the contribution to the driving energy for convection from secular cooling of the Earth's interior, including core-mantle differentiation, is uncertain. The detailed pattern of the convection flow is thought to be highly sensitive to the viscosity of the Earth's mantle and to its spatial variations. The history of mantle convection is closely linked not only to the history of plate motions, but also to the removal of heat from the Earth's interior and to the chemical evolution of the crust and mantle.

Gravitational and Magnetic Fields

The longer wavelength variations of the geoid and gravity field provide information on the density distribution in the mantle. Since these density inhomogeneities drive mantle convection, the measurements can be used to infer the structure of mantle convection. The interpretation of the long-wavelength features of the gravity field will be complemented by improvements in seismic resolution of density variations. Complete upper mantle coverage will be available from surface wave tomography. Lower mantle heterogeneity can be determined with more complete coverage and with an average resolution of about 200 km. To achieve this resolution worldwide requires a much denser global distribution of digital seismometers, including sea floor deployment. The long-wavelength part of the geoid shows a high degree of correlation with the lower mantle seismic heterogeneities. The inferred relationship between density and velocity places constraints on the viscosity structure of the mantle and the resulting relief on the core-mantle boundary. The intermediate wavelength part of the geoid correlates with the distribution of slabs and upper mantle velocity variations, constraining the density variations in these regions. At shorter wavelengths, lithospheric contributions, together with inherent limitations in seismological resolution, will make the interpretation much more patchy. There will still be uncertainty as to the relative contributions of convective or elastic support of geoid features. Understanding of the energetics of mantle convection will probably continue to be limited by ambiguities in interpretation of heat flow data. High-resolution global maps of heat flow will never be available, but surface-wave tomography shows a high degree of correlation with heat flow.

The direct inference of long-term (i.e., post-glacial) variation from gravity measurements began in 1983, with a determination of changes in long-wavelength harmonics. The geoid is not static! Estimates of changes in higher zonal harmonics can be expected by 1995, but determination of tesseral harmonic trends seems unfeasible. Determination of tidal effects on satellite orbits will be refined, and will help solve the problem of tidal dissipation. It also can be expected that the static gravity field will contribute to the understanding of post-glacial rebound. A notable recent achievement, the determination of the time variation of J_2—the oblateness—helps resolve the viscosity of the lower mantle.

By 1995 the oceanic geoid should have an uncertainty of less than a meter and a spatial resolution of 10 to 20 km. The principal means to this resolution will be the DOD satellite Geosat; the task group assumes that its results will become available for scientific publication. Because of the variations introduced by ocean processes, improvements in knowledge of the ocean geoid will be slow, dependent on more and more complex analyses of growing data sets. Knowledge of the geoid over land areas, however, is much more variable. In developed accessible areas, surface measurements provide low levels of uncertainty and good spatial resolution. Nevertheless, surface data are not now available in many areas because of either physical or political inaccessibility.

By 1995 it is also hoped that the Geopotential Research Mission (GRM) will provide gravity data over the continents with an accuracy of 2 mgal and a spatial resolution of 100 km. Because of the lower limit on spacecraft altitudes it is not possible to significantly improve this resolution from satellites.

The core interacts with the mantle in two important ways: it transfers heat into the base of the mantle, and it exerts torque on the mantle. The former contributes to and may even drive thermal convection in the deep mantle, and the latter causes changes in the length of day and in the orientation of the Earth's axis of rotation in space.

Although it is widely accepted that the Earth's magnetic field is maintained against dissipative ohmic decay by self-excited dynamo action in the liquid outer core, the details of the process remain obscure. It is uncertain whether there is (1) thermal convection driven by radioactive heating distributed throughout the outer core, or (2) slurry convection near the top of the core, or (3) chemical convection driven by compositional change and latent heat release at the boundary between the liquid outer core and the more solid inner core. We do not know (1) whether the core dynamo is laminar or turbulent, (2) whether there is a weak toroidal field, or (3) whether the toroidal field strongly dominates the poloidal field. Can the core magnetic field really change globally within an interval of less than 2 years as appears to have been the case during the "geomagnetic impulse of 1970?" Are such jerks rare or common, and how large can they be?

Clearly, light would be shed on many of these questions if we could obtain data necessary to construct an acceptable model for the fluid motion beneath the core-mantle boundary (CMB).

Probing magnetically more deeply into the fluid core does not yet appear to be feasible, but the construction of "synoptic weather maps" for the fluid near the CMB is technically realizable. The pattern of that motion could reveal the type of convection going on, and help determine the strength of the toroidal field, which is attenuated to unobservable levels at the Earth's surface. If the fluid motion is steady in time, it is uniquely determined by the time-varying vertical component of the magnetic field at the top of the core. Currently, the major scientific challenge for the subject of core fluid dynamics is to develop sound methods for extracting horizontal fluid motions near the top of the core from magnetic measurements taken at and above the Earth's surface.

Magsat resolved crustal magnetic anomalies and obtained an excellent snapshot of the main magnetic field, but gained hardly any useful instantaneous information on secular variation. Nevertheless, Magsat data were recently compared with observations at earlier epochs to determine, magnetically, the depth of the CMB supporting the frozen-flux model of the core and the nearly insulating model of the mantle.

Geomagnetic secular variation, crucial for studies of core dynamics, is currently best measured by ground-based permanent magnetic observatories and repeat stations, which are sparsely and very unevenly distributed over the Earth's surface. They will continue to play an important supporting role, especially during the long intervals between magnetic main field missions, but they cannot provide an adequate data base for global studies on their own.

The GRM will be of substantial importance for studies of core dynamics. Its low-altitude, carefully monitored, circular polar orbit will provide a significant improvement in our ability to resolve the crustal magnetic anomalies, which must then be removed from the data to expose the main field emanating from the core. The comparison between Magsat and GRM magnetic anomalies should go far toward establishing their repeatability, stability in time, and spatial scale of variation. However, snapshots of the main field at intervals of a decade and more cannot teach us about the continuous time evolution of the magnetic field, including the possible existence of short-term magnetic impulses.

The Earth's outer core has long been thought to be a relatively homogeneous molten body. Seismic tomography studies of

the lower mantle indicate the existence of mantle density variations, which are matched by long-wavelength features of the geoid
and which require kilometer-scale relief on the core-mantle boundary to support the mantle density variations. Relief at the top
of the core may play a controlling role in core dynamo mechan-
ics, and changes at the core-mantle boundary may control such
phenomena as westward drift of the geomagnetic field. Tomography can potentially be used to observe topographic variations in
the boundary of the fluid outer core and variations in the core.
Such observation will contribute to understanding the thermal behavior of the core, chemical differentiation occurring there, and
eventually the operation of the magnetic dynamo, one of the most
significant and least understood processes of the Earth.

The non-dipole terms of planetary magnetic fields extrapo-
lated down to the generating regions are appreciable. These regional variations must be significant in the overall behavior, particularly in reversals of dipole polarity evidenced by the Earth's
remanent magnetism. Magnetic observations obtained over the
last 150 years indicate a rate of change of the magnetic field such
that the non-dipole terms would appear quite different in a few
thousand years. This so-called secular variation can be inferred by
magnetometers on appropriate satellites—ideally, small dedicated
spacecraft orbiting for decades at altitudes of 1000 km or more.
The Magsat satellite launched in 1978 established a baseline for
measuring long-term changes. However, the orbit was far from optimum for this purpose, and hence estimates of the field generated
by the core are affected not only by solar-wind-induced variations,
but also by the remanent magnetism of the crust. Furthermore,
the duration of the mission was much too short to obtain any
estimate of temporal variations in the field.

The GRM will improve the resolution in determination of
variations in remanent magnetism to about 100 km. An Explorer
satellite with a magnetometer should greatly improve estimates of
secular variations to about the tenth degree of harmonics.

For an integrated approach to core fluid dynamics, we require
long-term, nearly continuous vector magnetic data from nearly
circular, polar orbits at sufficient height to minimize data contam-
ination by crustal anomalies and ionospheric currents. It would
probably suffice to turn on a satellite vector magnetometer for a
week or two every 6 months, but long overall mission duration

is vital. Magnetic signals diffuse downward through the conducting mantle rather slowly, so long time spans of surface data are
also required to probe deeply for the mantle conductivity profile.
The Magnetic Field Explorer mission, currently under discussion,
would be ideal for this purpose. It would be especially useful if
it were in orbit during the period of the GRM, for it could then
supply the excellent baseline main field model above which the
GRM crustal anomalies stand out.

Structure of the Earth's Interior

It is a good first approximation to assume that the Earth's
structure is radially symmetric. However, mantle convection in-
evitably entails lateral heterogeneity. Recently, inversions of seismic travel times have been used to obtain mantle heterogeneities
on a global scale. These "tomographic" studies can also be used
to measure the topography of the core-mantle boundary.

First interpretations of the tomographic results indicate the
importance of these studies toward understanding the dynamics
of the Earth. The results for the upper mantle show that the
anomalies under the mid-ocean ridges vary significantly in their
depth extent. Anomalies are associated with hotspots, shields,
and back-arc basins and may provide important information on
their origin.

An illustration of the implications of this new geophysical approach is the effort to interpret large-scale seismic velocity anomalies in terms of variations in density. The seismological results
therefore can be integrated with interpretation of the gravity field
and even the magnetic field (roughness of the core-mantle boundary is related to the westward drift of the non-dipole field), and
can provide constraints on the rheology. Elements of the pattern
of the flow in the mantle can be predicted and can be used to
constrain the range of lateral variations in temperature and composition. While some progress can be gained through refinements
in analysis of the existing data base, it is clear that *a major improvement in the resolution can only be achieved by a significant
increase in quality and quantity of seismic observations with a new
global seismic network.*

Another important application of such a global seismographic
network is studying earthquakes. Quantitative determinations of
the energetics and kinematics of earthquake sources are applied

now systematically to several hundred events per year. Accumulation of such data for the past 7 or 8 years allows us for the first time to monitor variations in the pattern of stress accumulation and release. In particular, indications of stress migration and diffusion have been inferred for several subduction zones.

Several interesting properties can be indirectly deduced from measurement of variations of Earth's orientation. Measurement of Earth's precession constant already provides the most accurate estimate of Earth's moment of inertia. Recent results from the Polaris very long baseline interferometry (VLBI) network indicate that the amplitude of the annual nutation differs from model predictions based on a hydrostatic Earth. The datum may indicate that the core-mantle boundary has a small non-hydrostatic ellipticity on the order of 300 m in amplitude that is somehow supported by mantle convection. Detection of the free core rotation (namely, its frequency) or changes in other nutation constants (semiannual and 18-year) could corroborate this interpretation. These measurements call for a commitment to a long-term observation program utilizing Polaris, lunar laser ranging, and the Laser Geodynamic Satellites (LAGEOS).

Another major structural parameter that may be inferred from these kinds of studies is inner core/fluid core density contrast. One method would be to attempt to observe the translational modes of the inner core by measuring the short-period variations in gravity at Antarctica. So far, this experiment has failed to observe any signal. An alternative would be to detect the inner core free wobble through its effect on the mantle wobble. The core wobble frequency is proportional to the core ellipticity and density contrast and requires core rigidity for time scales of at least a few years' duration. The existence of this mode requires an excitation mechanism (core dynamo?). The detection of this mode may, therefore, provide new information related to dynamo processes at depth. Detection of the effect of inner core on mantle nutation is another possibility, although this must be separated from the effect of core-mantle ellipticity.

Data from laser ranging to reflectors placed on the Moon during the Apollo missions have been collected on a routine basis since 1969. Precise ranges have been obtained from Texas and Hawaii. This data set has proved extremely useful in determining variations in earth rotation and polar motion, tidal recession of the Moon, possible detection of a lunar core, and detection of lunar

Chandler wobble. More operational ranging stations have recently been added in France and Hawaii. A refitted Australian system will soon be added to this list. The expectation is that these stations will range more frequently, with more accurate ranging systems, to obtain 5-cm or better normal point accuracy. The promise is that this system will complement the Polaris VLBI network in estimating earth orientation variations in earth rotation, polar motion, and nutations. More precise measurements of lunar parameters are also expected from this growing network.

Currently, VLBI and other techniques are greatly improving constraints on the rate of the Earth's rotation and the direction of the rotation axis (wobble). At the opposite end of the spectrum, we finally have two-digit accuracy on the rate of tidal dissipation. It can be expected that techniques of instrumentation and analysis will continue to improve over the coming decade. The NAVSTAR satellites of the Global Positioning System (GPS), a carefully located set of strong sources, will make a significant contribution. By 1995 there should be an order-of-magnitude improvement in the sorting out of the contributions of the atmosphere, tides, and core-mantle interaction to the spectra of rotation and polar wobble, and we may have the first reliable determination of a change in the pole path due to an earthquake.

History of Earth's Crust

The origin and early evolution of Earth might be considered so remote in time that there is no hope of gaining any meaningful information from measurements made today. However, the Earth has memories on various time scales and ancient rocks do exist. There are also objects in the solar system of various ages and at different stages of development that provide information complementary to earth-derived data.

What constraints do we have? The most obvious constraint is the age and composition of the oldest rocks. These rocks show that water was present on the surface of the Earth and a magnetic field existed at about 3.8 eons BP. The presence of ultramafic komatiites in the Archaean suggests that the mantle was hotter at that time than it is now, or else that it was easier for high-density magmas to reach the surface. Isotopic data show that the Earth was separated into chemically distinct reservoirs in its early history. The decay of the radioactive isotopes of potassium, thorium, and uranium means that heat production in the mantle was a factor of 3 or 4 higher in the Archaean than at present, with the consequence that the surface thermal boundary layer was thinner and melting temperatures were probably achieved at shallower as well as greater depths. Continental crust formation was rapid in the Archaean and appears to become episodic and generally less intense as we approach the present.

The fact that primordial gases are still emerging from the mantle means that the Earth is not completely outgassed, although this observation alone does not allow us to constrain the amount of outgassing. It seems clear that the gross differentiation of the Earth into atmosphere, hydrosphere, crust, mantle, and core occurred prior to the beginning of the geological record.

The study of meteorites and other objects in the solar system places constraints on the solar nebula and on the evolution of various-sized bodies that are relevant to the formation and early history of the Earth. The recognition that melting was widespread on small objects in the early solar system and that earth-sized planets were able to melt or vaporize incoming objects during most of the growth phase, all point toward a hot origin. The density difference between melts and refractory phases probably resulted in a chemically stratified body—stratification that was occurring while the planet was growing by a process akin to zone refining.

The presence of thick, buoyant anorthosite crust and KREEP in the Moon is best interpreted in terms of a magma ocean. A similar situation on the Earth would lead to a dense eclogite layer, the high-pressure equivalent of basalt. A thick, cold crust is impossible on an earth-size planet because of pressure-induced phase changes. The bottom of a thick crust is denser than "normal" mantle. In fact, the eclogitization of thick crust may be the instability that caused the early geological record to be erased.

An understanding of the other terrestrial planets is of obvious importance, since they are all at different stages in their evolution because of differences in size and surface temperature. The Moon and Mars represent thermally old bodies because of their small size, but since they are relatively inactive they retain surface evidence of ancient processes.

The chemical composition of the mantle and of its various regions is also of obvious importance. This can be approached by modeling the seismic velocities in terms of chemistry and mineralogy. This in turn requires laboratory data regarding the physical properties of mantle minerals at high pressure and temperature.

THE EARTH'S AIR AND WATER

The atmosphere, ocean, land, and biota form an interactive system that determines the current state and evolution of the portions of the Earth in which life evolved and on which we now live. A detailed understanding of this interactive system is necessary for prediction of climate on time scales ranging from months to centuries. A study of the Earth's atmosphere, hydrosphere, and cryosphere is an essential step toward gaining this understanding.

Atmosphere

In order to understand the physics, chemistry, and dynamics of the atmosphere, one needs an accurate description of its state not only all around the globe, but repeatedly and at a variety of time scales. The progress in this field of science has therefore largely been paced by our capability to monitor the various atmospheric parameters globally, rapidly, and repeatedly.

Ever since the launch of the first meteorological satellite in 1960, followed by dozens of research and operational satellites over the last 25 years, we have made substantial progress in accurately describing both the troposphere and the stratosphere. On the very short time scale—hours to days—the fluxes of mass, momentum, and energy between the land, the ocean, the ice, the atmosphere, and the biota, all proceed rapidly and are important in forcing day-to-day changes in global weather patterns. These fluxes are modulated by the daily cycle and latitudinal gradient of the solar heating, the eddies in the ocean currents, evaporation of water from the ocean and land, the rotation of the Earth, and the distribution of land masses and their topography. The transfer processes are initiated on the microscale, driven by the turbulent motions of the boundary, and may end up, only a day or two later, as a major contribution to a thunderstorm or a violent blizzard up to 1000 miles away. The predictability of weather depends on understanding the process of surface exchange and the internal atmospheric processes. More specifically, it depends on how well we can model the system, how realistic the model input parameters are, and whether we can practically integrate such a model on a computer.

In this context the global observational system composed of satellite and ground networks now provides us hourly to daily measurements of the state of the global atmosphere. Such information includes cloud cover, cloud heights, water vapor, and atmospheric temperatures, and these data are now used routinely for weather forecasting purposes around the world. The observations have been found to be helpful in tracking severe storms, including hurricanes over the oceans, and successfully predicting the time and place of their arrival over land. Data on the state of the atmosphere are also fed into the numerical models of the medium-range (several days to weeks) weather forecasts, but so far the degree of improvement in the accuracy of these predictions has remained controversial. It is possible that the dynamics of the atmosphere at the mesoscale level are basically unpredictable beyond 10 days, the small currently unmeasured microscale instabilities growing into macroscale perturbations in about 2 weeks. However, through recent experimentation on the global scale it appears that if accurate measurement of tropospheric winds, precipitation, and fluxes of water vapor from the surface into the atmosphere are made available, our ability to forecast weather for a period of a week to 10 days may improve substantially. In the early 1990s we will be attempting to measure these rather elusive parameters, allowing us to test their impact on weather predictability.

Substantial progress has also been made over the last decade in documenting the changing chemistry of the troposphere. The amount of carbon dioxide in the atmosphere has increased from 315 ppm in 1958 to more than 345 ppm in 1985. Ozone is observed to vary on a wide variety of time scales. It has also been found that the amount of methane in the atmosphere has increased by more than 10 percent in the last 10 years (see Figure 2.1). The chlorofluoromethanes (freons) are currently increasing globally at from 1.7 to 6.2 percent per year, depending upon the chemical species (see Figure 2.2). Nitrous oxide is increasing at a rate similar to that of carbon dioxide. These changes in concentrations are of great concern because these gases contribute to the greenhouse effect and therefore may increase the temperature of the surface of the Earth.

By 1995 we expect to have established the precise trends of these trace gases in the troposphere by extensive ground and space

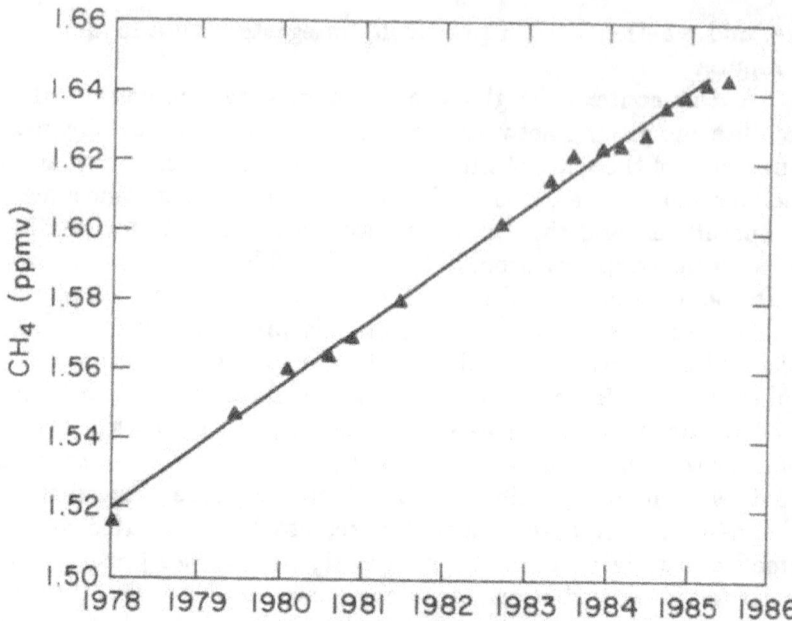

FIGURE 2.1 Globally averaged methane concentrations in surface air from January 1978 to March 1987 as measured by D.R. Blake and F.S. Rowland. Air samples were collected simultaneously from northern and southern latitudes.

SOURCE: Printed by permission from Donald R. Blake and F. Sherwood Rowland, *Journal of Atmospheric Chemistry*, volume 4, pages 43-62 (1986), plus unpublished data continuing the measurement series.

experimentation. However, any prediction of how these gases will build up in the atmosphere in the future depends largely on the rate of their cycling through the ocean and the biosphere. Such a prediction therefore will have to await a more comprehensive understanding of the total interactive system. It now appears that the cycling of carbon, nitrogen, sulfur, and potassium through the atmosphere, biosphere, and oceans is interlinked, and that the tropospheric concentrations of molecules like methane, carbon monoxide, nitrogen oxides, ozone, and sulfur dioxide are interrelated by complex chemistry involving the OH radical, which has so far remained unmeasurable.

For the troposphere, the main avenue of research so far has been related to the physical processes forced by the interaction of radiation with the atmosphere and the energy released through evaporation and condensation of water. In the next 10 years there will be a new emphasis on study of the implications of chemical changes in the atmosphere on climate and the hydrological cycle. The next step after that will be the incorporation of the biota and the deep oceans into the total system, leading toward an understanding of global change on the time scale of decades to centuries.

The stratosphere is the 30- to 40-km-thick region of the atmosphere above the tropopause (6 to 16 km) and is dominated, at least energetically, by the presence of the ozone layer, which shields the surface of the Earth from harmful solar ultraviolet radiation. The stratosphere also has an important influence on climate at the Earth's surface. Stratospheric aerosols, for example, play a role in the global energy budget. Most of them appear to be derived from sulfur gases (sulfur dioxide and COS) transported up from the troposphere continually and by the episodic injections of volcanoes such as Mt. Agung and El Chichon. Because of low temperature and presence of water vapor, once the gases arrive in the stratosphere they condense into small droplets of sulfuric acid and persist for up to several years. Through the solar energy absorbed by ozone and aerosols in the stratosphere and the role these constituents play in adding to the greenhouse effect in the troposphere, it is clear that the stratosphere plays a substantial role in modulating the energetics and dynamics of the lower atmosphere. The Upper Atmosphere Research Satellite (UARS), scheduled to be launched in 1991, will address radiative exchanges, chemistry of the trace gases, and circulation in the stratosphere. The principal questions remaining in 1991 will be the role of change in total ozone and the coupling of the stratosphere with the troposphere and the surface. This latter problem will have to be tackled by innovative and simultaneous measurements of physical, chemical, and radiative parameters in both the stratosphere and troposphere. Only then will we be able to make any measurable progress on such issues as, for example, whether the solar variability has any influence on tropospheric weather and climate.

The mesosphere and lower thermosphere, extending between 70 and 200 km, are the least explored regions of the Earth's atmosphere. They are influenced by varying solar ultraviolet and gamma-ray radiation, particle precipitation, electric and magnetic

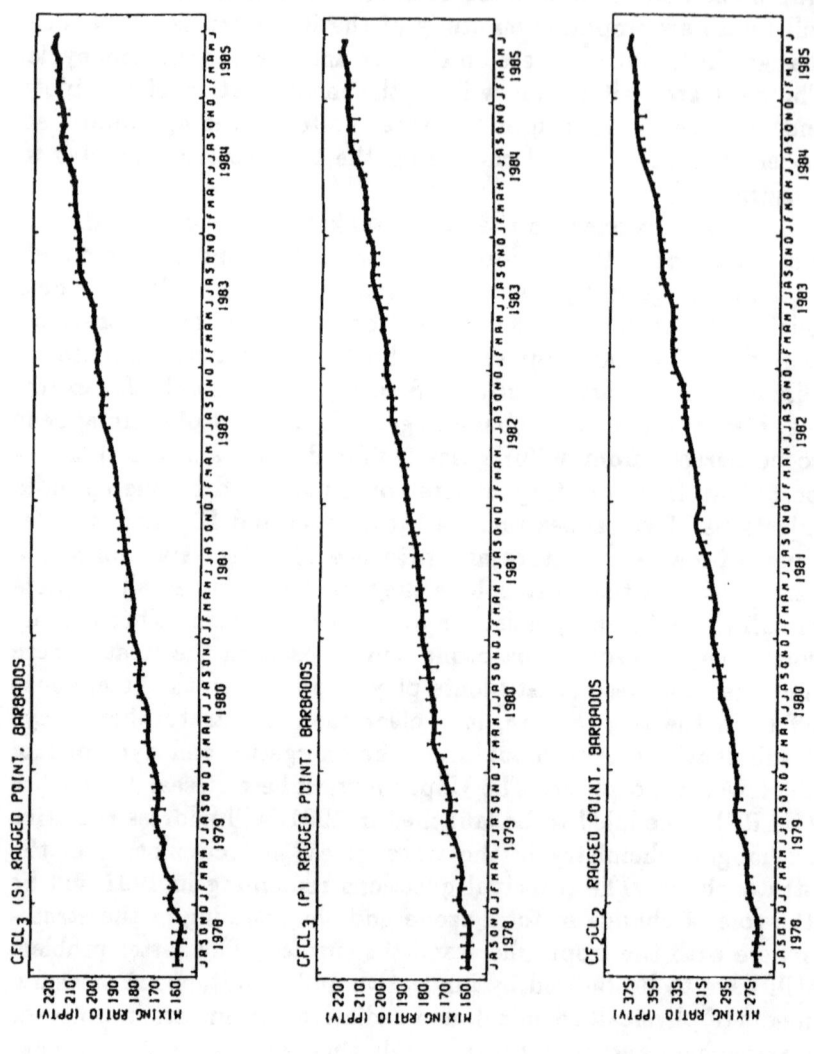

33

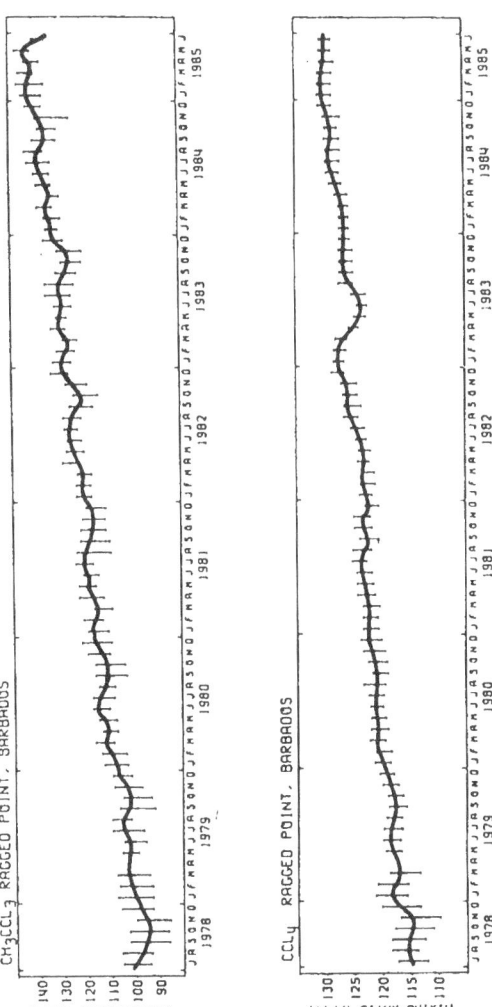

FIGURE 2.2 Monthly mean mixing ratios of the trace gases trichlorofluoromethane, dichlorodifluoromethane, 1,1,1-trichloroethane, tetrachloromethane, and nitrous oxide measured over the period 1978-1985 at the Barbados station of the Global Atmospheric Gases Experiment (GAGE). These four species are seen to be increasing at rates of 4.9, 5.2, 6.2, and 1.7 percent per year, respectively, at this site. Qualitatively similar results are obtained from the other GAGE stations located in Tasmania, Samoa, Oregon, and Ireland. All of the trace gases shown are important in the chemistry of the ozone layer and in the greenhouse effect.

SOURCE: Prinn et al., private communication (1987).

fields, and upward-propagating waves from the lower atmosphere. This atmospheric region absorbs and globally redistributes energy and momentum deposited in the atmosphere during episodic events like geomagnetic storms, solar flares, and solar proton outbursts. It is not known just how deep into the Earth's atmosphere the chemical, radiational, and dynamic effects caused by solar variability and such episodic events penetrate. This atmospheric region has an important influence that is not currently understood on the transmission, reflection, and absorption of waves and tides propagating upward from the lower atmosphere. The few available measurements and model studies of the lower thermospheric region indicate that it is a dynamically active region with significant coupling between chemistry, radiation, and dynamics, and with charged species in the ionosphere. The atmospheric region between 70 and 200 km plays an important role in buffering solar-terrestrial interactions.

At the present time, the fields of global thermospheric, mesospheric, and ionospheric dynamics are healthy, with a good balance between theory, numerical modeling, and observations. As we approach 1995, the observational program looks much worse because there are no opportunities for satellite programs to obtain measurements of the dynamics of the upper atmosphere and ionosphere, and its response to solar and auroral activity.

Oceans

Earth is a water planet with two-thirds of its surface covered by life-sustaining oceans. If there were no oceans, there would be no efficient formation of carbonates and essentially all carbon dioxide would be in the atmosphere. Without the oceans and ocean currents, which redistribute heat around the globe, climatic extremes would be far more severe. Processes at work in the ocean are also important in the budgets of many chemically and radiatively important gases, such as carbon dioxide, nitrous oxide, methane, and many sulfur-containing gases. In the polar regions the ocean interacts with the atmosphere to produce sea ice, which is itself a sensitive indicator of global warming or cooling and which controls the rate at which heat can escape from the polar oceans into the overlying atmosphere.

Both world weather and long-term climate change are strongly

linked to ocean behavior. In the past 20 years, our ability to forecast weather over several-day periods has significantly improved
with the routine availability of observations from meteorological satellites. However, it is increasingly apparent that further
improvement—particularly in prediction over weeks and seasons—
requires an improved knowledge of ocean behavior.

The Joint Oceanographic Institutions satellite report noted
that "new technology that provides global views of the oceans
using satellite-borne instruments, coupled with new high-speed
computers, promise major breakthroughs in our description and
understanding of the ocean. The data from satellites have shown
that we can vastly improve our understanding of ocean processes.

"Two recent events have emphasized the oceans' importance
in global climate: the disastrous 1982-83 El Niño, which caused
billions of dollars in damage and loss of fish resources as well as
considerable loss of life, and the potentially harmful effects of increasing carbon dioxide levels in the atmosphere from the burning
of fossil fuels. We now know that our ability to understand and ul-
timately to predict events associated with a severe El Niño or with
the warming predicted from a carbon dioxide increase is severely
limited by a lack of ocean measurements. New global information
available from satellites, coupled with data from the interior of
the ocean, can meet this need. Existing weather satellites operated by NOAA, the Department of Defense, and non-U.S. space
agencies provide some routine ocean surface observations, but we
still lack crucial data on surface winds, ocean currents, biological
productivity, and the gravity field of the Earth."

What do we expect to have measured and learned about the
ocean and its interaction with the atmosphere, solid Earth, and
biosphere by the year 1995 with the planned satellite and in situ
programs? Plans for oceanography from space for the decade
prior to 1995 involve the flight of four new missions addressing the
circulation and biology of the world's oceans, collection of data
from ongoing operational satellites, and major field programs that
will use the satellite data as a global integrating element to study
processes in situ. By 1995 we also expect to have additional global
synoptic descriptions of the ocean. This will be coupled with new
models of the ocean that accurately describe its turbulent physics
on next-generation computers. The fundamental questions of the
processes that drive the circulation and mixing, of the processes
that are responsible for sustaining ecosystems, and of long-term

interactions of the ocean with the geology at the bottom and at the coasts will be partially answered.

Even with this large injection of new resources into the system, at the end of the measurement period in roughly 1995 we can expect to have only a 3- to 5-year snapshot of ocean processes, which in fact are energetic over a wide range of time scales. Figure 2.3 shows the energy density as a function of period for the ocean, and the time scales to be addressed by the above missions. Moreover, although we will have measurements of wind and radiation as drivers for the ocean, we will not have measurements of precipitation over the ocean. Thus the data and studies up to 1995 will reveal only a part of what we need to know to develop predictive models of the overall system.

A simple example will illustrate this point. The El Niño climate anomaly is a major perturbation on the coupled ocean-atmosphere system. Yet in even a 20-year period, we can expect to see only three or four of these events. Even with a 20-year data set, it is not likely we would be able to have sufficient data to begin to understand this complex problem of coupled turbulent fluids on the rotating Earth. To understand the effect of increasing carbon dioxide on the systems will take an even longer time series. High-deposition-rate sediments may provide the longer time record.

Fresh Water and Ice

Part II of the CES strategy addressed the issue of fresh water and ice in some detail. The report observed that "water is the most abundant single substance contributing to the global biomass and cycling through the biosphere. It is also a vital resource that in many countries is extensively managed for delivery to agricultural lands. Nevertheless, less than 0.1 percent of the water on Earth is directly usable, and only half of that is easily accessible near the Earth's surface. The global supply of usable water is fragile: the global inventory of surface fresh water can be drained by the natural processes of evapotranspiration and runoff within a few years. Despite some local water surplus, many regions of the world either now or soon will experience water shortages because of inadequate quantity or quality.

"Water in its frozen state also plays important roles in global and regional energy budgets, in weather and climate, and in the

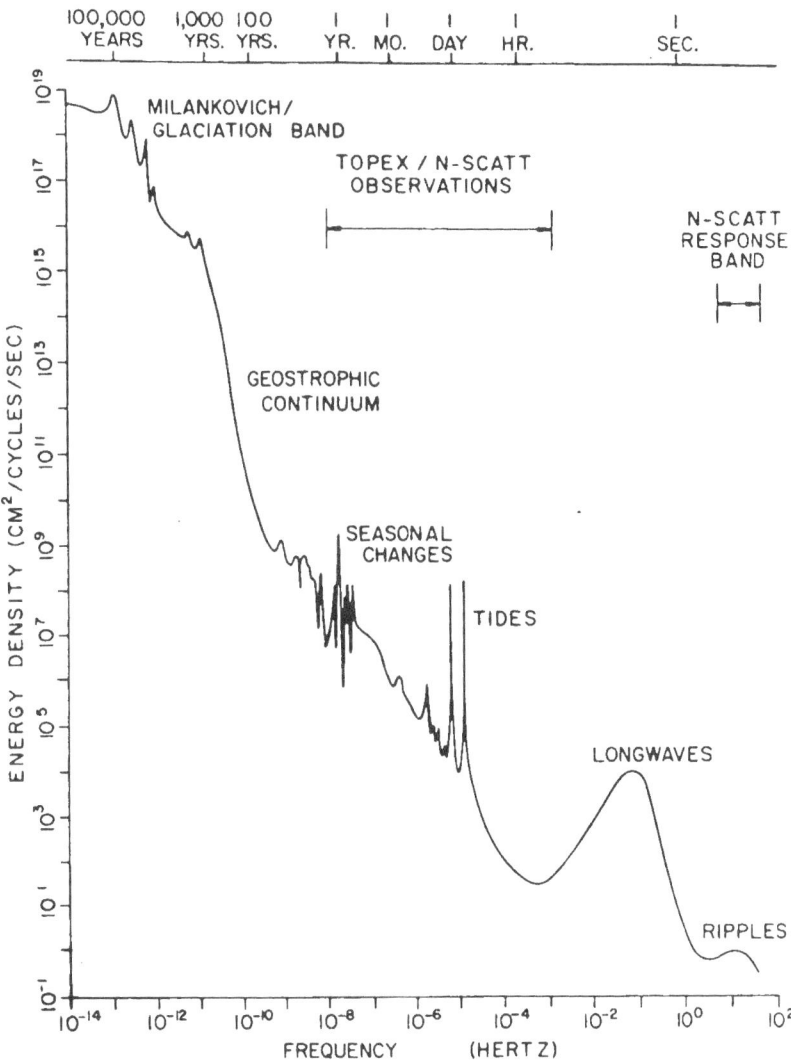

FIGURE 2.3 Schematic frequency spectrum of ocean dynamics/kinematics with sea level as variable.

SOURCE: C. Wunsch, Massachusetts Institute of Technology, 1986.

annual supply of surface and ground water. Snow, glaciers (including ice sheets and ice shelves), ice in seas, lakes and rivers, and ice in the ground are the major components of the frozen part of the global system—the cryosphere. Each of these components possesses its own distinctive physical and chemical properties; seasonal and geographical variations; mechanisms of formation, movement, and loss; and interactions with the atmosphere, ocean, and land surface. Over the past 30 years our knowledge of the cryosphere has grown steadily with exploration being pursued for both scientific and economic motives. We now know that ice and snow cover play a very important interactive role in the dynamics of the Earth's climate and that, properly monitored, the world's ice volume provides a sensitive indicator of climatic change.

"Theoretical studies have shown that both anthropogenic change (e.g., atmospheric carbon dioxide increases) and natural perturbations (e.g., volcanic dust) may have significant influences on the magnitude and distribution of global precipitation and temperature. Since the latter influences evapotranspiration, then runoff (approximated as the difference between precipitation and evapotranspiration) will be highly sensitive to anticipated global climatic change. Changes in river flow and lake volumes, accelerated soil erosion, and desertification all represent significant and measurable hydrologic responses to climatic changes operating on time scales of several years to decades. These changes all lag the climatic forcing factors by differing amounts, and they occur on top of natural variations that must be understood to adequately manage their immense consequence for food productivity and other human uses of the Earth. Likewise, the global hydrologic consequences of large-scale water diversion projects, urbanization, tropical deforestation, and regional irrigation projects need to be assessed.

"Certain aspects of global snow and ice research are feasible only through observations from space. First, the remoteness and scale of the cryosphere make real-time, global-scale data acquisition possible in practical terms only with satellite-borne sensors. The vastness of the cryosphere is not always appreciated: glaciers alone cover 11 percent of the Earth's land surface; over 50 percent of the land is covered (and uncovered) by snow each year; and sea-ice covers 12 percent of the world's ocean. If the area where icebergs are commonly encountered is also considered, ice is observed over 22 percent of the ocean's surface. Second, many fundamental snow and ice investigations require the collection of data at regular intervals without any limitation imposed by clouds, inclement weather, or darkness; again, the necessary regular observations are in most cases possible only from Earth-orbiting satellites."

The Challenge: Climate Prediction

The issues involving the Earth's air and water spheres that will most likely still be unresolved in the 1990s are those that possess complex and synergistic interaction between land, ocean, and atmosphere; those that require global, long-term, and nearly simultaneous observations of a multitude of parameters; and those that need direct involvement of scientists from a number of traditional disciplines. These are the central issues of predicting climate in the future.

On short time scales (months to seasons) it is now becoming clear that the tropical oceans play a very large role in determining the behavior of the global atmosphere. However, it is also known that volcanoes can suddenly inject large quantities of aerosols, sulfur, and other elements into the atmosphere and perturb the climate on a large scale as well. The recent events of El Niño and El Chichon are excellent examples of these two phenomena occurring simultaneously, followed by record perturbations in climate around the world for the next two years. To deconvolve the mechanisms and the cause and effect relationships is a problem that spans across the disciplines of oceanography, meteorology, and atmospheric chemistry, and therefore will not be resolved by studying any one of them in isolation.

On a little longer time scale (years to decades) there is the problem of increasing carbon dioxide and its impact on the climate of the future. A study of this problem involves understanding the global carbon dioxide cycle, which immediately brings us to the changing biomass of the world, the nutrients in the oceans, and the biogeochemical cycles in general. Also, the process of changing biomass impacts the surface albedo and evapotranspiration rates and therefore can affect the climate directly. We now recognize that, in addition to carbon dioxide, there are many other gases in the atmosphere (methane, freons) that are increasing globally and may have an impact on climate by their own greenhouse effect.

In an even longer time frame (thousands to millions of years) the problem of climate variability is linked to orbital changes of the Earth, to deep ocean circulation, and to continental drift arising from plate tectonics. The record of these climatic changes is largely contained in sediments, ice cores, and the morphology of the rocks. Explanation of these variations is one possible test of any theory of contemporary climate change. To understand the mechanism we need information, for example, on deep ocean water, plate motion, and vegetation-climate feedbacks. Again, the need to cut across discipline lines to make any progress on these issues is self-evident.

Mars also apparently had warmings and coolings. If the record on Mars can be compared with the Earth, it will test the idea that climate changes are driven by variations in solar output.

LIFE ON EARTH

Our objective is to characterize the interactions between the biota and other components of the Earth—most notably the atmosphere, oceans, and the solid Earth. These data will be used both to make predictions about the short-term future behavior of climate and selected ecosystems, and to better understand the past behavior of the biosphere and climate system, particularly as this relates to changes in the physical and chemical environment of planet Earth.

The importance of biosphere-geosphere interactions has been clearly recognized in the National Research Council report, *Global Change in the Geosphere-Biosphere*. This task group endorses the conclusions reached there, recognizing the urgency of the International Geosphere-Biosphere Program in the context of current concerns about large-scale changes in the environment wrought by humankind. The task group wishes to emphasize, however, that the study of biosphere-geosphere interactions must be placed in the larger context of a Mission to Planet Earth. The unique state and history of the Earth is connected inextricably to the fact that life originated on this planet. Life has evolved interactively with the physical and chemical environment ever since.

Biological perspectives on this research agenda are discussed by the Task Group on Life Sciences and presented in a separate volume. The Task Group on Earth Sciences emphasizes that these two discussions are complementary and point toward a common set of technological requirements and intellectual goals.

Relation of Physical and Biological Earth History

From the perspective of the solar system, the Earth appears to be unique in its capacity to sustain life. Not only does the Earth sustain life today, it has done so for at least the last 3.5 billion years. Organically preserved microfossils found in the oldest known unmetamorphosed sedimentary rocks document the early evolution of bacterial communities containing morphologically diverse organisms. Stromatolites (the sedimentary-free fossils of microbial mat communities) further indicate that some early organisms were phototactic, and stable carbon isotope ratios in kerogen strongly suggest a carbon cycle driven by photosynthesis. Indeed, the evidence available from paleontology, geochemistry, and microbiology suggests that anaerobic biogeochemical cycles were well established by the time the Earth's oldest surviving sedimentary rocks were deposited.

Metabolism links biology closely to atmospheric science. Organisms both produce and consume gases, thereby affecting the composition of the atmosphere. Not only is atmospheric oxygen related to the evolution of cyanobacteria and (later) algae and green plants, but many other important gases such as nitrous oxide (a product of bacterial ammoniac oxidation and nitrate reduction) and methane (a product of bacterial methanogenesis) owe their presence to biogenic processes that track back to the Archean diversification of bacteria. Evolutionary innovations in biological structure and physiology may have had profound effects on the Earth's radiation balance, climate, rates of surface weathering and erosion, and the rates of deposition, diagnosis, and distribution of sediments. The extent to which our planet's surface, hydrosphere, and atmosphere have been altered by life throughout its history is a scientific problem of major theoretical and practical significance.

Global Biota: Revelations from Space

Nearly 4 billion years of evolution have produced a diverse biota estimated to include as many as 10 million distinct species. These species are not distributed randomly across the planetary surface; rather, they are organized into an ecological hierarchy based on biological interactions of species with similar physical tolerances. Local, recurrent associations of species form communities that interact with the physical environment to form ecosystems. Congruent ecosystems can be grouped in bigger units called biomes, characterized by similarities in plant growth forms, community structure, and productivity, among other things. Some biomes, such as the Great Plains grasslands, have a high capacity for primary production and so are crucial to the support of the human population. Others, such as the Arctic tundra biome, are less productive, but are believed to contain large quantities of organic carbon in their soils and may be sensitive indicators of global changes in temperature or pollution levels. The areal extent of biomes, gas fluxes into and out of them, and their mean primary productivity are very imprecisely known at present; however, because biomes have spectral properties that permit identification and analysis by remote sensing, the perspective from space makes possible the detailed global analysis of biome distribution and production.

Global ecosystem analysis, especially of agriculturally important systems, will provide information prerequisite to the scientific estimation of the Earth's capacity to support the growing human population. Analysis of forested biomes, especially the tropical rain forests that are increasingly being destroyed, is essential for efforts to understand the changing carbon dioxide content of the atmosphere. There can be no accurate quantitative models of biogeochemical cycles and their interactions until such an ecosystem or biome inventory has been made. Once such data are produced, a host of hitherto insoluble geochemical and biological problems will be brought within our grasp. It is clear that we must continue to monitor the state of the Earth in this context for the indefinite future, to document changes in biotic regimes as they occur, and to aid in the development of quantitative models to assess the impact and origin of changes observed.

Biogeochemical Cycles

The state of knowledge and the crucial significance of water and the hydrological cycle have been discussed. But the movement of material through living organisms involves many more elements. The carbon, nitrogen, phosphorus, and sulfur cycles, to name just a few, are critical to the mechanisms and maintenance of life on Earth. The state of knowledge and major problem areas for each of these four major cycles are discussed below as a prelude to determining a strategy for studies in the 1995 to 2015 period.

The task group emphasizes mainly the shorter time scales. However, the role of plate tectonics—subduction and volcanism—in circulating and remobilizing these constituents may be crucial to the maintenance of life on Earth on the longer time scales. It is essential in this context to understand the nature of the forces responsible for volcanism for the recycling of critical elements such as carbon. To what extent are the changes in climate characteristic of past ages of our planet attributable to fluctuations in rates of cycling of carbon and associated variations in atmospheric carbon dioxide? The Mission to Planet Earth, with its interdisciplinary focus, must seek to provide answers to these questions.

According to the NRC report *Global Change in the Geosphere-Biosphere*, "there is abundant evidence for change at present. Most obvious perhaps are changes in the composition of the atmosphere—of CO_2, CH_4, CO, N_2O, NO_x, SO_x and O_3—and changes in the chemistry of precipitation. There are more subtle effects associated with altering practices of land and energy use, and of waste disposal. Anthropogenic changes are superimposed on natural fluctuations, and it is difficult to separate the anthropogenic from the natural changes that are taking place today. There are clues, however, from the record of the past.

"An impressive body of information has accumulated recently to suggest that fluctuations in CO_2 may have played an important role in regulating at least some of the major changes in climate of the past. The level of CO_2 was approximately 200 ppm during the last Ice Age. It rose by about 50 percent, to approximately its present value, in only a few thousand years, 10,000 to 12,000 years ago, ushering in the present interglacial period.

"We can reconstruct the history of CO_2 back to about 60,000 years before the present using air trapped in bubbles in ancient ice preserved in Greenland and in Antarctica. A more indirect technique, based on analysis of the isotopic composition of carbon in the carbonate skeletons of marine organisms in ocean sediments, has allowed us to extend the record even further, to about 400,000 years ago. The correlation with climate is striking. High CO_2 is invariably associated with warm conditions, low CO_2 with cold; and indeed, changes in CO_2 appear to precede changes in climate.

"Carbon dioxide is but one of several gases with the potential to raise the temperature of the Earth. Infrared radiation from the planetary surface is also absorbed and reradiated by methane (CH_4), nitrous oxide (N_2O), and O_3 and by the industrial halocarbons, CF_2Cl_2 and $CFCl_3$. On a molecule-per-molecule basis, these gases are much more efficient than CO_2 in altering the radiative balance of the present Earth, and their concentrations are also changing. Their cumulative effect on climate over the past several decades may be comparable with that of CO_2."

The Carbon Cycle

As was pointed out in Part II of the CES strategy: "There are two central chemical processes in the carbon cycle: aerobic oxidation and anaerobic oxidation. Increases in the rate of aerobic oxidation are the probable cause of the observed increases in atmospheric CO_2; increases in the rate of anaerobic oxidation may be the cause of the observed buildup of CH_4. The case of CO_2 exemplifies many of the limitations in our current understanding of global cycles as well as important gaps in current data sets. . . .

"The possible effects of human interference with the natural cycle of carbon by burning fossil fuels, harvesting forests, and converting land to agriculture are reflected most clearly by the phenomenon of increasing concentration of atmospheric CO_2 [see Figure 2.4]. If current trends continue, the atmospheric concentration will exceed 600 parts per million by volume by the year 2040—more than 2 times the preindustrial level. The increase in CO_2 is important because, in contrast to atmospheric O_2 and N_2, CO_2 absorbs infrared radiation emitted by the Earth and prevents the escape of some of the normally outgoing radiation. This is known as the 'greenhouse' effect, a phenomenon operating with high efficiency on Venus.

"At present, our ability to interpret the carbon cycle and thus predict future CO_2 concentrations is confounded by unresolved imbalances in the carbon budget. Simply stated, the annual budget does not balance unless (1) fertilization effects, either terrestrial or aquatic, partly offset deforestation minus regrowth, (2) the imbalance diminishes from reductions in the estimate of the rate of deforestation or increases in the regrowth, (3) the oceanic uptake is underestimated, or (4) there are natural variations in the global rate of carbon uptake by the biota that are not yet recognized."

Methane is the second most abundant form of carbon in the atmosphere. Its presence reflects the importance of localized media

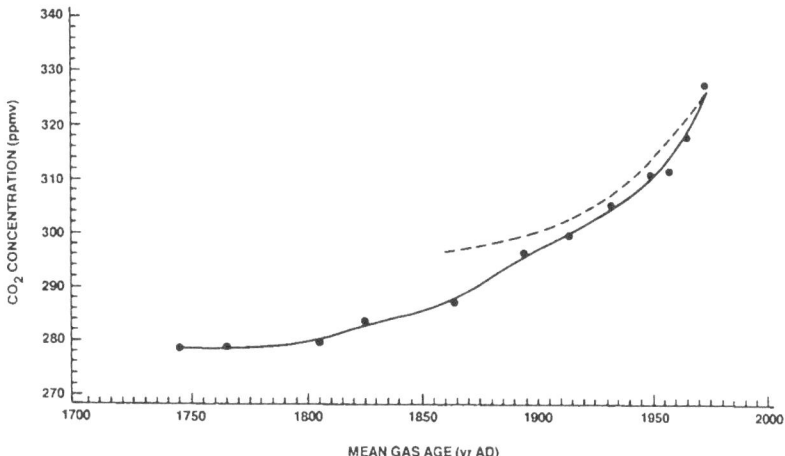

FIGURE 2.4   Measured mean CO_2 concentration plotted against the esti-mated mean gas age. The horizontal axis of the ellipses indicates the close-off time interval of 22 yr. The uncertainties of the concentration measurements are twice the standard deviation of the mean value, but not lower than the precision of 1 percent of the measurement. The dotted line represents the model-calculated back extrapolation of the atmospheric CO_2 concentration, assuming only CO_2 input from fossil fuel. Atmospheric CO_2 concentrations measured in glacier ice formed during the last 200 years calibrated against the Mauna Loa record for the youngest gas sample.

SOURCE: Neftel et al., *Nature*, volume 315, pages 45-57, 1985.

where oxygen is deficient, as in swamps and the soil of rice paddies, for example, or in the digestive tracts of ruminants and a variety of other animals, including termites. Its abundance is now increasing at a rate of about 2 percent per year. The concentration in the atmosphere appears to have doubled since the sixteenth century. Why? How will it vary in the future? What was its level and scale of variation in the past?

The Nitrogen Cycle

The NRC report *Global Change in the Geosphere-Biosphere* noted that "nitrogen occurring in compounds as single atoms (fixed nitrogen) is chemically versatile and essential for life, with a range of oxidation states from −3 to +5. Processes that break the N-N bond (nitrogen fixation) are relatively slow, amounting to less than 0.2×10^{15} g/yr of N. Recombination of fixed nitrogen to form N_2 is also slow, owing largely to the kinetic stability of inorganic, fixed nitrogen (NH_4^+, NO_2^-, NO_3^-) in solution. The recombination reaction is carried out biologically by bacteria using NO_3^- and NO_2^- as electron acceptors (denitrification). Denitrification takes place in anoxic, organic-rich locations such as flooded soils and estuarial sediments, bottom waters of some deep ocean basins and trenches, and in low-oxygen or anoxic waters at intermediate depths in coastal upwelling regions. Denitrification is essential to the preservation of the present level of atmospheric N_2. In the absence of biological processes, the atmospheric nitrogen cycle would be open, leading to accumulation of NO_2^- and NO_3^- in the oceans. It is unclear how the global system acts to establish a balance between fixation and denitrification. Mechanisms directly coupling nitrogen fixation to denitrification have not been identified, and indirect connections are not obvious.

"Nitrogen is cycled through the biosphere at rates 10 to 100 times as large as the rate for fixation of N_2. Inorganic fixed nitrogen (NH_4^+, NO_2^-, NO_3^-) is assimilated into terrestrial biomass at a rate of about 3×10^{15} g/yr of N, but this influx is balanced by decay of organic material. The rate at which inorganic fixed nitrogen is consumed and recycled by biota in the oceans is roughly 2×10^{15} g/yr of N, with a large uncertainty. Internal cycles of mineral and organic nitrogen are essential links in the life-support system of the planet."

The supply and distribution of fixed nitrogen thus affect not only the biosphere's productivity, but also the chemical and radiational environments for life. Changes in the abundances of atmospheric nitrous oxide and nitrogen oxides attest to the importance of contemporary changes in the biogeochemistry of nitrogen. How did levels and distributions of fixed nitrogen vary in the past and how did they relate or act to influence the climatic and biospheric condition of the planet?

The Phosphorus Cycle

Part II of the CES strategy pointed out that "phosphorus is an essential element for life. It is relatively abundant in the crust of the Earth, but it exists principally as insoluble minerals (apatite, iron phosphates) or as absorbed phosphate. These forms are not available for biological uptake, and consequently, phosphorus is often a limiting nutrient in soils, lakes, and perhaps even marine systems. Atmospheric transfer processes are unimportant for phosphorus, in contrast to carbon, nitrogen, and sulfur. Rather, the major phosphorus exchanges are associated with dissolved and particulate transport in rivers, and with weathering processes and diagensis in soils and sediments. There are thus important connections between the hydrologic cycle and the phosphorus cycle.

"Most of the phosphorus in rivers is insoluble and biologically unavailable, and there are major questions about the actual fraction of river-borne phosphorus that manages to participate in the biological cycle and the time scale for effective transfer from rivers to the oceans. . . . Additional uncertainty is associated with storage of phosphorus in estuarine and coastal sediments."

This latter issue is important since this phosphorus could be mobilized during epochs of low sea level (e.g., during glaciation) and delivered to the ocean where it could be responsible for an increase in biological productivity and a consequent drop in carbon dioxide. What determines the level of oceanic phosphorus? How has it varied in the past? What factors are responsible for change in oceanic phosphorus and what are their consequences?

The Sulfur Cycle

Sulfur is also an essential element for life, but unlike nitrogen and phosphorus it is rarely limiting. It exists, like nitrogen, in a variety of oxidation states, from -2 in sulfides to $+6$ in sulfates, and is cycled among these states by the biota, by volcanoes, by combustion of fossil fuels, and by atmospheric reactions.

As noted in *Global Change in the Geosphere-Biosphere*, "sulfur enters the atmosphere in two dominant ways. Combustion of fossil fuels adds sulfur in the form of SO_2. Microorganisms in soils and in the surface waters of the ocean putatively contribute additional amounts in the form of $(CH_3)_2S$, H_2S, and other reduced sulfur gases, but the precise amount is unknown and controversial. These reduced gases are oxidized to SO_2 on time scales of hours to days. Anthropogenic and natural inputs of SO_2 to the atmosphere are apparently comparable in amount.

"The SO_2 is oxidized to sulfate and in this way removed from the atmosphere on time scales of several days. The sulfur oxidization processes depend on atmospheric levels of the OH radical and thus on the abundances of atmospheric O_3, H_2O, nitrogen oxides, and hydrocarbons.

"In the soil and in the ocean photic zone, sulfate is taken up by plants and microorganisms. The sulfur is then recycled to the atmosphere through processes of decay; some accumulates in organic matter in ocean sediments. On geologic time scales, sedimentary sulfur is returned to the ocean-atmosphere system through volcanism."

In addition, Part II of the CES strategy states that "for the sulfur cycle, there is a need to identify and quantify the anthropogenic and biological fluxes of reduced sulfur gases and determine whether these fluxes are subject to change. A far better understanding of the atmospheric chemistry of the reduced sulfur gases and the SO_2 from combustion is also needed. Of particular concern in this chemistry are the roles of heterogeneous reactions, the coupling to atmospheric nitrogen and carbon chemistry, and the mechanisms for dry and wet deposition. Finally, we require far more information on the manner in which sulfuric acid deposition affects the biology and geochemistry of terrestrial ecosystems."

The external information needed to model these processes includes the major biological sources and sinks of organic carbon and active nitrogen, and inputs of sulfur and other compounds from volcanic activity. Urban pollution is a topic all by itself, but is a major regional source of tropospheric ozone, oxides of nitrogen, sulfur dioxide, and other ingredients of larger-scale problems like acid rain. Also required is information on the transport and mixing capacity of the atmosphere. Clouds play a key role in catalyzing certain reactions and in scavenging water-soluble products in precipitation. Although in principle the atmospheric transports and cloud fields are available as part of the modeling and data base of the physical climate system, in practice, considerable additional effort is required to make them useful for chemical purposes. Just as important are the internal measurements, which give guidance as to which chemical processes are most significant, and give confidence that they are being modeled correctly.

Human Activities

There are concerns about both (1) human effects on the environment, and (2) effects of natural phenomena on man. Both are exacerbated by growing population, and industrial and agricultural development. The basic scientific issues of biogeochemical interaction have been discussed above; the global issues and necessary measurements will be discussed in Chapter 3.

PLANET EARTH IN THE SOLAR SYSTEM

A global understanding of the Earth entails explanation of why it is different from other planets. These differences arise from a relatively few fundamental properties—mass, composition, distance from the Sun, rotation rate—but often the secondary manifestations are greater than expected. Furthermore, the other planets must be considered in any meaningful consideration of the Earth's formation, which constitutes the starting conditions for the Earth's evolution.

At least eight bodies in the solar system—seven planets and one satellite—have significant atmospheres whose dynamics and chemistry should be explained by any comprehensive theory of atmospheres. Dynamically, the atmosphere of Mars is most similar to the Earth's, in that it is a relatively thin fluid envelope around a rapidly rotating rocky planet, subject to marked seasonal variations because of an appreciable tilt of the rotation axis to the orbit axis. Mars, like the Earth, has cyclonic systems of weather and, on a much longer time scale, appears to have undergone glacial waxing and waning. But beyond confirming a few fundamentals, Mars does not contribute significantly to the solution of problems regarding the Earth's atmosphere, most notably because it lacks an ocean. Compositionally, the atmospheres of both Venus and Mars could have been quite similar to the Earth's with one striking exception: the much greater complement of primordial inert gases in Venus, a difference that must be a consequence of circumstances of formation 4.5 billion years ago. Otherwise, the examples of Venus and Mars act as strong constraints on theories of atmospheric evolution: any worthwhile theory must account for the loss of water from the Venusian atmosphere (most likely by photodissociation, leading to loss of hydrogen and trapping of the free oxygen in surface rocks), resulting in the development of the greenhouse effect, so that the comparable portion of carbon dioxide in Venus stays in the atmosphere rather than being incorporated in the ocean and thence in carbonate rocks.

Pioneer Venus revealed that the high surface temperature of

Venus arising from its massive carbon dioxide atmosphere has had important consequences for the solid planet. Venus undoubtedly has mantle convection, since it would have incorporated almost the same energy sources as the Earth. But its rocky surface is quite lacking in indicators of plate tectonics, such as an interconnected ridge system like the ocean rises on Earth. Hence the boundary layer of the mantle convection within Venus must be at depth, below a basaltic and sialic crust. The entire surface of Venus may be covered by a continent-like crust; the Pioneer altimetry indicates that there is only one predominant level of topography, rather than two, as on Earth.

Venus is also significantly different in that it has no magnetic field. It certainly differentiated an iron core, but if the pressure is too low and the temperature too high no inner core will form, thus eliminating a possible energy source for a geodynamo. These marked differences of the planet most similar to the Earth in size and composition act as important constraints on models of the early evolution of the Earth including crustal formation, outgassing, and other events of the early Archean, more than 2.5 billion years ago.

Constraints of a somewhat different sort in understanding the Earth arise from consideration of its formation. Moon rocks and meteorites indicate, by their radiological ages, that formation of all the planets took place within a few 10 million years some 4.57 billion years ago. The retention of abundant hydrogen and helium by Jupiter indicates that it was quite massive before the formation of the terrestrial planets was markedly advanced. Hence, the dynamical circumstances of terrestrial planet formation were dominated by the gravitational influence of Jupiter, which probably was important in inducing growth to only four planets plus a satellite, rather than a larger number of smaller bodies. This growth pattern probably led to the terminal stages of formation being characterized by a few great impacts. Important evidence of that includes the differences in rotation rates and inert gas retention between Venus and the Earth, and the anomalously low iron content of the Moon. It is the current consensus that the Earth was probably hit by a very large body—perhaps bigger than Mars—which led to the lifting off of the material that made the Moon, and which removed virtually all of any primordial atmosphere from the Earth. Consequently, the Earth formed very hot, leading to early core separation and outgassing of the atmosphere and ocean. A likely by-product of that was formation of a crust that was similar to the Moon's. Another consequence of the hot beginnings of the Earth was obliteration of any evidence of this crust. The lunar crust is 10 percent of the mass of the body. By contrast the terrestrial crust is less than 0.4 percent.

Another application of comparative planetology to Earth history is the record of cratering on the surfaces of the Moon, Mars, and Mercury. This record indicates that throughout this history there has been a sporadically declining infall of bodies, a few sizable enough to have global effects catastrophic for major parts of the biosphere. Firm chemical evidence of such effects has only recently been deduced: most notably, the marked iridium spike at the horizon marking the end of the Cretaceous period 60 million years ago.

The history of the Earth shares many common threads with the histories of one or more of the other inner planets, including early global differentiation of crust and core, outgassing and evolution of the atmosphere, early bombardment of the surface by a heavy flux of meteoroids, and development of a global magnetic field and magnetosphere. The Earth has many attributes not shared, however, with any other known planet, including its oceans, the oxidized state of its atmosphere, its tectonic plate motions and the consequent complex history of crustal deformation, and its life forms. A continuing challenge to the earth and planetary sciences is to account for the profoundly unique attributes of the Earth in the context of the common processes that have shaped the formation and evolution of the solar system.

Various attempts have been made to use the Earth as a detector of cosmic, stellar, and solar system events. In order to do this, however, the Earth itself must be better understood.

Earth is a collector of extraterrestrial particles and thus can be used to estimate the current meteorite flux. Some have used extinctions throughout the geologic record to propose periodicity, or at least episodicity, in the influx rate of larger objects. Meteorites falling to Earth are one guide to processes in the early solar system and processes in small disrupted objects.

3
The Earth as a System—
A Global Perspective for Future Planning

INTRODUCTION—OBJECTIVES AND GRAND THEMES

While rapid progress is being made in several of the earth sciences, outstanding problems will remain in 1995 and persist indefinitely. In this chapter the task group presents a synthesis of objectives, based on four grand themes: the surface, crust, and interior of the Earth; the atmosphere and oceans, including the hydrologic cycle; the biosphere; and the impact of mankind.

These themes provide a philosophical basis for the necessary measurements and experiments over the years 1995 to 2015. The themes arise from the fact that *in order to understand the integrated functioning of the Earth as a system it will be necessary to abandon the conventional subdivisions of earth science for an integrated study of processes.* There are elements of each of the traditional disciplines involved in understanding the major questions regarding the Earth, its origin, evolution, structure, and present operation. There are also elements in common in the measurement programs required to address these issues.

The complex of scientific issues discussed in earlier sections can be used to establish a base for a coherent plan of action. The task group draws upon these statements of progress and problems to attempt a synthesis through the identification of grand themes that encompass the many threads of scientific investigation of the Earth. The themes are mostly concerned with changes and interactions, which implies that we must have an understanding of the baselines. Clearly, the Earth is not a steady-state system, and must be viewed as evolving. This evolution can be seen as an ongoing process, where the basis of our extrapolation—both forward and backward—is the present. The necessity for extrapolation has widely differing time scales and reliabilities for different parts of Earth. The evolution of the Earth can also be viewed as a comprehensive process, starting from its formation out of the solar nebula and leading eventually to a state of stable stratification as internal energy sources run down (as they have on the Moon). This view of the Earth must depend to an appreciable degree on a comparative planetological approach.

Conventional wisdom presents the Earth as a roughly steady-state system, with oscillations about its mean state and occasional wild excursions. Nearly all human activities implicitly assume this steady state. But a cursory examination of the historical record indicates we are on a one-time binge of a couple of centuries with respect to population and petroleum, and perhaps in other areas of comparable importance, such as arable soil. Longer time scales are associated with oscillations in climate (10^4 to 10^6 years), and in the solid Earth (10^6 to 10^9 years). These characteristics are oversimplifications; the real Earth has appreciable oscillations, both endogenic and exogenic, over a wide range of time scales. Most striking are catastrophic events, such as volcanic eruptions, asteroid impacts, and earthquakes. The geologic record indicates that on a million-year time scale events occur that are thousands of times as energetic as the Mount St. Helens outbursts, with short-term global consequences for the climate.

Climate variation on the 100,000-year time scale is dominated by the waxing and waning of glaciers. Currently the Earth is enjoying an unusually warm period. The temperature variation inferred from oxygen isotopes of deep-sea cores appears to be correlated with variations in the Earth's orbit, but to have appreciable non-linear enhancement. This problem may be solvable with global observation of the temporal variations of the Earth's albedo, sea surface temperature, and other relevant parameters in response to the milder seasonal variations in our time.

Study of Venus indicates that the present Earth may be radically different from its early environment. It is essential to study early Earth—to determine whether its atmosphere was reducing. When did the oxidizing environment occur? Did it coincide with the development of life? Did this change coincide with the onset of plate tectonism? Are these connected in time? Are they cause and effect?

The contemporary behavior of the solid Earth is also anomalous in that there are now an exceptional number of continents compared to that typical in the Phanerozoic (the last 600 million years). Geologic evidence also indicates significant variations of plate tectonic rates and patterns on time scales of 10 to 100 million years. These oscillations are evidence of the mantle dynamic system and its interaction with the lithosphere. Again, understanding must be advanced by observations of the present state plus extrapolation based on theoretical modeling and the geological record. Imposed on this general evolution of the natural Earth is the rapidly expanding effect of man on the landscape. Better understanding of this impact is a major scientific interest as well as a matter of great practical concern. The themes identified below grow out of the need for such understanding.

Necessity drives earth scientists to ask for a variety of measurements—simultaneous, continuous, and on a worldwide basis—from the obvious global tool, artificial satellites. Because satellites in orbit are external to the Earth, the answers they give are incomplete and must be supplemented by measurements at closer range or in situ, by laboratory experiments, and by theoretical modeling. Our discussion below identifies the global issues for each grand theme, and then specifies the measurements required.

GRAND THEME 1: STRUCTURE, EVOLUTION, AND DYNAMICS OF THE EARTH'S INTERIOR AND CRUST

Global Issues

About 70 percent of the Earth's mass is mantle, the rocky region between the crust and core. A leading problem of solid earth science can be described as mantle climatology: the description of variations in composition and physical properties of regions of the mantle, how these heterogeneities relate to the dynamics, and the resulting evolution over the eons. The crust is very much dependent on the dynamics and evolution of the mantle. The crust is a region of importance greatly out of proportion to its mass since it is the intermediary of the solid Earth for the several interactions with the hydrosphere, atmosphere, and biosphere. Somewhat more separate is the core, mostly fluid, whose principal manifestation is the Earth's magnetic field.

The geoid and detailed seismic imaging show that the mantle is inhomogeneous, both radially and laterally. Geochemical data indicate that there are ancient reservoirs in the mantle, but their locations and relation to the seismic inhomogeneities are unknown. The geophysical and geochemical data must constrain the style of mantle convection and contribute to the understanding of earth evolution and the nature of the energy sources.

The lithospheric plates are the cold, surface boundary layers of mantle convection cells, but several aspects of mantle convection are poorly understood. These include the energy source—primordial heat or radioactivity—and its distribution.

Many plate boundary phenomena are also ill understood; most important are those associated with subduction—how subducted material produces magma and how this magma rises to the surface. Furthermore, the nature of subduction zones varies greatly, apparently influenced by the natures of both the overlying and subducted materials: oceanic-under-oceanic (e.g., Tonga), oceanic-under-continental (e.g., the Andes), and continental-under-continental (e.g., the Himalayas).

Continents evidently grow by the accretion of island arcs, but it is unknown whether they grew predominantly by this process in the past. The stabilization of continental crust and lithosphere, and the control by crustal thickness are the consequence of mantle-crust interactions not yet well identified. Mass balance calculations based on isotopic data require appreciable recycling of crustal material; the proportion recycled by subduction of sediments, delamination of lower crust, or other mechanisms is also as yet unknown. The extent to which continental basalts and the associated upper mantle arise from a reservoir distinct from the source of mid-ocean ridge basalts needs to be inferred much more precisely.

Planetary magnetic fields measured to date show a wide range of behaviors, plausibly arising from major differences in fundamental characteristics of the planets. However, these plausibilities are not yet proven and, as in most complicated problems, the solution is to be found in the examination of details. Fundamental to the strong magnetic fields of the Earth, Jupiter, and Saturn (and the nonfield of Venus) are planetary dynamos: interactions of convection, electromagnetic induction, and rotational dynamics occurring in fluid interiors of high electrical conductivity. Problems that are likely to persist beyond 1995 are the energy sources for these dynamos, the scales and patterns of the motion of fluids, the temporal evolution of the flow, the boundary layer interaction with overlying material, and the values of key physical properties. The Earth offers the best opportunities to observe details relevant to these processes. Pursuit of this branch of earth science requires a combination of new satellite data to be used in conjunction with those existing from previous systems, plus ground-based observations. The two quantities of great interest and importance are the vertical magnetic field and its first time derivative, or its continuous time dependence, measured everywhere over the Earth's surface.

The Measurements Required

Structure and Chemistry

A number of specific measurements are required to describe, in three dimensions, the variation of physical parameters and chemical/mineralogical composition at all depths within the Earth's interior. These include global seismic wave propagation studies to describe lateral heterogeneities up to at least spherical harmonic degree and order 20. Regional seismic wave propagation measurements are also required to provide detailed images of major features such as subduction zones, fine structure of the crust and lithosphere, and selected areas near the core-mantle boundary. Smart ground stations, portable seismic stations, and ocean bottom systems will be needed for these measurements.

In addition, gravity observations (global and regional), geologic mapping using space techniques, geochemical and petrological analyses, and high-pressure, high-temperature laboratory experiments to understand the properties of terrestrial materials under these conditions will be necessary.

Dynamics

To understand mantle convection and the resulting motion and deformation of the surface plates, it is necessary to study the dynamics of the Earth's interior. Although it is generally accepted that the lithospheric plates are the cold, near-surface boundary layers of mantle convection cells, many aspects of mantle convection are still uncertain.

Interactions between plates are responsible for a large fraction of the Earth's seismicity, volcanism, and mountain building. Many of the fundamental processes are poorly understood. Episodic accumulation and release of stress at plate boundaries are responsible for great earthquakes. Lines of volcanoes are generally associated with subduction. But what happens to subducted material to produce magma? How does this magma rise to the surface? Why are some plate boundaries broad, such as in the western United States and China, and why are others relatively narrow, as in the Andes? What processes are responsible for the elevation of major mountain belts?

With nearly real-time transmission of data through satellites, seismologists are now prepared to derive advanced quantitative models of faulting within an hour after an event has occurred. Exercise of such a capability has clear implications for society and for science. Post-seismic rebound instrumentation, for example, can be rapidly deployed to a hypocentral region. The Global Positioning System will greatly facilitate these projects and will be a key element in these studies.

A global array of digital seismometers and geodetic devices telemetering via satellites to central "observatories" is the solid Earth equivalent of a versatile, multispectral telescope or a large-aperture radio telescope. The inside of the Earth is now a candidate for imaging just as are other objects in the universe.

An important aspect of the study of the solid Earth is its rheology, commonly parameterized as viscosity. It is significant for problems ranging from mantle convection to wobble and polar wander. Radial variations of viscosity are poorly constrained; lateral variations, while necessarily large because of temperature variations, are virtually unquantified. The correlation of seismic tomographic and geoid data with the heat flow from the Earth makes it possible to place bounds on the viscosity variations.

Specific measurements of the surface gravity field and geoid are required to provide information on the interior density distribution within the Earth. Satellites have provided a large fraction of our current data base. At present, our primary need is improved data over remote mountain areas such as the Andes and Himalayas.

Full understanding of the geoid requires seismic studies of the interior.

A Magnetic Satellite Mission dedicated to measuring the long-wavelength (400 km or more) components over several years—and preferably decades—would be particularly useful. The satellite could be rather simple; its orbit should be polar and about 1000 km in altitude, to assure both long lifetime and sensitivity to the long wavelength of the field. Substantial progress in understanding the origin of the Earth's magnetic field can be expected once we have detailed maps of the velocity and density variations in the outer core and in "region D," the mantle-core transition region.

In addition, seismic studies and seismic tomography can provide detailed three-dimensional images of the Earth's interior. These studies require measurements of travel times and the free oscillations of the Earth. They require a wide distribution of digital surface seismographs. Anisotropy, related to flow directions, can also be measured. Geodetic observations at the centimeter level could provide a wealth of information on tectonic displacements. A direct measurement of the plate motions would be obtained, and active tectonic processes could be studied in detail.

A number of other measurements are also needed. These include: electromagnetic measurements (satellite studies of the time variability of the electromagnetic field can be used to obtain the distribution of electrical conductivity within the Earth's mantle, electrical conductivity being a sensitive measure of the temperature within the mantle); ground deformation measurements (GPS, corner reflectors, readily deployable strainmeters); space mapping; space and ground chemical analyses; and measurements from aircraft and balloons.

Geological Mapping

Geological maps are perhaps the most fundamental data set in solid earth geoscience. The spatial distribution of rock types, when added to chronological and compositional data, allows detailed reconstruction of the geological evolution of a region. Only by mapping all of the continents to a uniform resolution can the record of the evolution of the Earth from 3.8 billion years be established. Thus, the importance of accurate geological maps cannot be overemphasized. Key operations for understanding the tectonic history of a region often center on the rates and magnitudes of processes such as faulting and uplift. Good geological maps afford an opportunity to compare estimates of short-term rates derived from geophysical techniques to long-term geological rates.

Geological maps are also critical in that they supply constraints to models. For example, it is important to develop models that relate strain buildup, fault slip, and earthquake occurrence to rheological properties of the crust and lithosphere. The spatial distribution of fault planes and the width over which shear is distributed across a fault zone are important parameters.

It is apparent that for many of the problems discussed above, highly detailed maps, coupled with extensive chronological data, are required. Such maps cannot be generated with space-based techniques alone, but require detailed ground investigations. Nevertheless, an important background data set, particularly for many poorly surveyed areas outside the United States, can be generated with remote sensing techniques. Both multispectral, optical-band stereo imagery, and synthetic and real aperture radar imagery can provide useful data for regional investigations.

Although many remote sensing data have already been generated by NASA, a surprisingly small amount has been used by the geological/geophysics community. The cost and time involved in acquiring and processing remote sensing data in its present form often make it prohibitively expensive for the average geologic mapping program. The time involved in searching the large variety of data archives also tends to limit accessibility. A centralized facility that would catalog, process, and make available such data to the geological/geophysical community would be an important step.

For many geological problems, spatial resolution of 10 m or better is required to adequately map the distribution of critical lithological units. Present space-based sensors are thus not adequate for many tectonic problems. Nevertheless, they can provide important constraints for regional problems and afford an opportunity to look at large terrains in a new, synoptic manner. Improved spatial resolution would greatly enhance applicability to other problems. The photographs of the Large Format Camera on the Shuttle have now established the extreme usefulness of 10-m resolution. There can now be no going back to 30-m resolution.

The present spectral resolution of the thematic mapper is a great improvement over other Landsat sensors. It nevertheless does not allow discrimination of most lithologic units. Higher spectral resolution, particularly in the infrared, is required to obtain even crude lithologic discrimination capability. Current coverage of thematic mapper imagery is, to some extent, limited by ground receiver capability, though this is expected to improve when another TDRS satellite becomes operational and as more ground receiving stations come on line. Present coverage with synthetic aperture radar imagery is extremely limited.

It cannot be emphasized enough that the strength of space-borne sensors lies in their global, synoptic coverage; hence, Shuttle deployment is of limited use. Global coverage is required to attack many of the significant problems in tectonics.

Global Topography

Global, digital topographic data are required for a number of geological and geophysical investigations. At present, data at adequate resolution are available only for the United States and a selected number of Western European countries. Topographic data are required for proper analysis of gravity data, in order to deconvolve the contribution of topography to a given gravity signal. More generally, analysis of coupled topography and gravity data allows the determination of subcrustal structure, gravity compensation models, and crustal rheological properties. Clearly, adequate topography data must be an integral part of any gravity mapping mission.

Sufficient topographic map coverage is lacking for many critical regions, including much of Africa, South America, and the Himalayas. Digital topographic data for the continents are useful for an astonishing range of purposes, including geophysics (for example, gravity compensation modeling), civil engineering (for site surveys), and botany (for example, species distribution and health, estimated from optical sensing techniques, as a function of altitude). It also has obvious applications in geology/geomorphology, and would aid remote sensing in general because registration of digital topography with other kinds of image data would allow correction of albedo effects and layover distribution in optical and radar data, respectively. Finally, altimetry data over the polar ice caps would allow calculation of ice-flow-driving stress and would aid in monitoring the long-term health of ice sheets.

Topography data with moderate resolution can be obtained economically with a dedicated Topographic Mapping Mission on the Space Shuttle. Global coverage can be obtained in three missions. The system would employ a microwave altimeter with a phased array antenna. The long dimensions of the antenna would generate a small footprint in the cross-track dimension (500 m) for the required spatial resolution. Electronic beam steering of the phased array would allow the appropriate swath width for complete global coverage. Synthetic aperture techniques would ensure adequate spatial resolution (500 m) in the along-track dimension. Real aperture techniques may allow the same coverage in one extended mission. Height resolution should be better than 5 m.

Higher resolution altimetry could be obtained with a scanning laser altimeter. Higher power requirements for such a system, in the range 2 to 5 kW, dictate deployment on a large permanent platform such as EOS. A pulsed laser with a pulse duration in the range 5 to 20 ns and a pulse repetition frequency in the range 2 to 4 kHz could generate global coverage in about 1 year with 100-m spatial resolution and 1-m height resolution. Technical improvements in the long-time reliability of lasers are needed for this purpose.

Surface Imaging and Sounding

Space-borne Synthetic Aperature Radar (SAR) systems have proven to be very useful for a variety of geological, botanical, and agricultural applications, as well as selected oceanographic and ice monitoring studies. Current generation space-borne SAR is restricted to single-frequency, single-polarization instruments. However, multifrequency and multipolarization capability and utility have been tested on aircraft, and are expected to be demonstrated before 1995 on the Space Shuttle with the SIR-C experiment.

Multifrequency radar can potentially be used to map parameters such as soil moisture, vegetation mass and health, and possibly the amount of snow pack. In arid regions, multifrequency SAR can be used effectively to distinguish and map shallow subsurface layers. Multipolarization capability at a given wavelength effectively maps volume-scattering properties. Perhaps the most obvious applications of multipolarization SAR are in the fields of botany and agriculture. Here, the orientation and volume density of plant leaves and stalks determine the relative proportions of backscattered energy at the various polarizations. Thus, multipolarization SAR can be used to map vegetation type and monitor vegetation health. A variety of imaging and sounding instruments on geosynchronous and polar platforms will be needed to obtain uniform global coverage.

GRAND THEME 2: ATMOSPHERE, OCEANS, CRYOSPHERE, AND HYDROLOGIC CYCLE

Global Issues

The central theme will be to establish and understand the structure, dynamics, and chemistry of the ocean, atmosphere, and cryosphere, and their interaction with the solid Earth, including climate, the hydrological cycle, and other biogeochemical cycles.

The Earth is unique in possessing an ocean and living organisms. There are growing realizations that the hydrosphere and biosphere, while constituting tiny fractions of the planet's mass, are crucial in establishing the character of the Earth in several ways.

The ocean, to a visitor from another planet interested in physics, would be most quickly recognized as the controller of water and heat, and the relative sluggishness of its circulation makes it the buffer to the variation of the atmosphere on time scales ranging from days to seasons. It also imposes its own pattern on decadal and longer time scales, as manifest in such phenomena as El Niño. The ocean and the cryosphere also are the main control on solar inputs to climate and weather. On longer time scales—10^2 to 10^6 years—the ocean, glaciers, and their distribution with respect to the land vie with volcanic inputs and solar variations in influencing climate. The relative roles of these different effects are still ill-understood; many observations remain that could improve our insight into these phenomena that are so important to human welfare.

It would be evident to a visitor interested in biology that the ocean would be essential to the development of life. Its margins have offered such stable riches as light, nutrients, perches, and protection from ultraviolet radiation through a reducing atmosphere. As life has evolved, its symbiosis with the ocean has made it a phenomenon covering the Earth's surface, as discussed below.

The influence of the ocean on the behavior of the solid Earth is important as well. It has a major effect on the chemistry of the continental crust through the intermediacy of its sediments. The hydrosphere may be important to island arc volcanism by fluxing magmatic activity in subduction zones. Just how hydrated sediments influence this process of continent-building is not clear, and has been much debated for decades. In addition, the ocean may significantly influence the mechanical behavior of the lithosphere; a relatively small proportion of water can weaken rocks so they are more easily subducted.

The ocean is also the most pervasive connecting medium for global biogeochemical cycles. The magnitudes of most chemical reservoirs and their rates of accumulation are strongly controlled by the ocean, which is significantly older than the ocean basins beneath it.

It has become apparent that the atmosphere, oceans, and the hydrologic cycle cannot be considered in isolation, but rather as a more complete system that includes interactions between the biosphere, solid Earth, and perturbations caused by solar variability and orbital changes. Many of the individual components of the system will have been investigated by 1995, and many of the techniques needed to address the Earth as a planet will have been developed.

The Measurements Required

In order to address this grand theme we will need to monitor and eventually understand the processes involved in global change of atmosphere, oceans, and their interaction with land. *We need long-term (on decadal time scales), consistent, and precise measurements of geophysical parameters such as the solar constant, stratospheric ozone, stratospheric temperature and aerosols, atmospheric trace compounds, surface albedo, land biomass, sea surface temperatures and topography, concentration of chlorophyll in the oceans, global cloudiness, and rainfall patterns and soil moisture.* Because the coverage has to be global and repetitive, space satellites are, in principle, ideally suited to provide these data consistently over time. Today a variety of satellites exist that are measuring some of these parameters routinely.

As we look beyond 1995, we see that the results from the 1985 to 1995 decade can be used to develop a cost-effective, long-term measurement scheme with a mix of satellite and in situ measurements. A program of space observation will therefore have to be designed that will provide unique global data sets made up of simultaneous observations of the atmosphere, land, and oceans for two principal purposes: to facilitate the setting of parameters for the various fluxes in the models, and to check the model predictions on a global scale. In order to achieve this, an observing program can be visualized that:

• Provides long-term and consistent data on some of the key parameters such as sea surface temperatures, ice cover, albedo, stratospheric ozone, and solar constant, so that we can begin to test the models at least on a decadal time scale.

• Develops new techniques for monitoring those parameters that are important in climate research but cannot be measured by the current space systems: rainfall, evapotranspiration, biomass.

• Assures the compatibility and continuity of some of the current observing systems: operational versus research satellites and U.S. versus non-U.S. satellites.

• Organizes field experiments that would help validate and authenticate the space observations.

At the same time, we will have to build a research community that is conversant with space technology and is drawn from a number of traditional disciplines of earth sciences such as volcanology, aeronomy, geology, oceanography, meteorology, glaciology, and biology so that a coherent attack on the climate predictability problem can eventually be mounted. The task group expects to see a continuation of the World Climate Research Program, which will be operating in earnest by the early 1990s, and to see the beginning of the International Geosphere-Biosphere Program. The latter will focus on interactions in physics, chemistry, and biology.

The major thrusts for atmospheric science beyond 1995 will involve:

• Development of a global measurement system for precipitation and evapotranspiration to define the latent heat budget for the atmosphere.

• Continuation of intensive studies of severe storms; their generation, steering, and dissipation.

• Development of a detailed understanding of the role of the biota in influencing the atmosphere—through trace gas uptake and emanations, through albedo influences, and through evapotranspiration—and the way in which these influences depend on environmental parameters.

For these purposes we will need an extensive program of in situ observations of processes in large land ecosystems, of tropospheric chemistry, of oceanic biogeochemistry, and of severe storms. Satellites will play a major role in precipitation measurements and complementary roles for severe storm, biota, and atmospheric chemistry investigations.

We expect that the most cost-effective program for oceanography will continue to be the relatively low-cost, single-purpose satellite missions that are properly intercalibrated. There is an important role for the Space Station, including polar platforms, in local and regional measurements that require high power for the sensors. Ground-based studies of high-deposition-rate pelagic sediments are also required.

The major science thrust for 1995 to 2015 will continue to be climate prediction for longer and longer time periods. As we move from interannual, El Niño-type events to long-term changes caused by increasing carbon dioxide, we must include the interactions of biology in the system. Understanding biology will be a major thrust for the 1990s and beyond.

To do this we will need a global satellite network together with major in situ programs to measure:

o *Ocean currents and mixing.* This includes a network of polar-orbiting satellites to measure sea surface topography, building on the results from the Ocean Topography Experiment (TOPEX) and the altimeters on the Earth Observing System (EOS). A larger in situ program, including moored and drifting stations, will be required to monitor mixing and sinking rates, as well as to validate the altimeter measurements and to measure currents below the surface.

o *Ocean-atmosphere interaction.* This includes a network of polar-orbiting satellites to measure sea surface topography and sea state, building on the ESA' Remote Sensing Satellite (ERS-1) and EOS results. In situ programs of moored and drifting stations again will be required to calibrate the satellite data.

o *Ocean chemistry.* New satellite techniques will most likely be available for monitoring ocean chemical parameters from space, especially salinity. These will be measured by multispectral techniques from polar-orbiting satellites, and must be calibrated by in situ measurements. In addition, chemistry measurement must be made in the bulk of the ocean by standard techniques to monitor long-term change.

•   *Precipitation and the hydrological cycle.* These are fundamental to the physical processes of climate and to the studies of climate variations. The flux of latent heat in the form of water vapor from the surface to the atmosphere, and its subsequent release through the condensation/precipitation process, constitutes the largest single heat source for the atmosphere.

Current rain gage networks on land are generally adequate to measure precipitation in heavily populated regions, but considerable standardization in worldwide observing and reporting practices is necessary. It is principally over unpopulated land areas and the oceans that precipitation data are lacking. Studies are under way to investigate measurements of rainfall over land through remote sensing via satellite.

The measurement of precipitation from space on a global scale is a formidable problem because as yet there are no methods that can be relied upon to perform under all circumstances around the world. Nevertheless, we already have some visible and infrared techniques that provide climatologically useful data. Also, over the oceans we are quite confident that by 1995 these methods can be extended by means of improved microwave radiometers. The use of combinations of measurement systems should be most valuable in filling the great gaps in our knowledge of oceanic precipitation, and it would serve to give us a better understanding of the sampling requirements and the adequacy of current surface observations.

The potential of space-borne radar as the ultimate tool for making direct precipitation measurements over the entire globe must be seriously considered. A number of approaches can be taken that involve conventional pulsed radar, coherent Doppler, dual wavelength, and polarization, among others. All these possibilities must be subjected to detailed feasibility studies. An important consideration is the possible combination of active and passive microwave techniques, and hybrid schemes involving visible and infrared channels. The goal is to overcome the long-standing obstacles to obtaining reliable global precipitation data.

In the area of climate research we will have to spend the next few decades improving global models in which atmosphere, land,

and ocean interact by exchanging energy, mass, and momentum on a variety of spatial and temporal scales. We will need data on fluxes at the boundaries rather than just on the state of the atmosphere or of the oceans. We emphasize again the role that ice-core and pelagic sediment studies can play in extending the record.

GRAND THEME 3: LIVING ORGANISMS AND THEIR INTERACTION WITH THE ENVIRONMENT

Global Issues

The overall goal for the study of global biogeochemical cycles is to improve understanding of the geologic, atmospheric, oceanic, and biotic reservoirs and their interactions in order to model and predict changes important to the biosphere and climate.

What must be known to permit us to understand the global balance of these cycles? Uncertainties in our understanding of the carbon cycles lead to serious difficulties in balancing the current budget of atmospheric carbon dioxide. There are a number of problems that must therefore be addressed: the extent of major terrestrial biomes and their carbon contents; the factors controlling the internal routes for uptake and release of carbon; the processes that control the exchange of carbon (both oxidized and reduced) between the interior, the atmosphere, biota, and oceans; and finally, the response of the carbon cycle to human perturbations.

While the amount of nitrogen fixation controlled by man annually is significant compared to natural fixation, it is still small compared with the existing fixed nitrogen pools in the soil and in the oceans. These pools therefore will be influenced only slowly. It will take at least several decades before significant global changes may be expected due to human activities; changes in particular localities, such as soil and water systems, may appear much sooner. But for the very reasons that it will be several decades before any significant global changes could be apparent, it will also take an equally long time for conditions to return to an earlier balance once a change is detected.

Specific issues are the following:

1. The elucidation of the storage and exchange of the principle elements in living things, in and between different components of the biosphere—the "biogeochemical cycles" of carbon, nitrogen,

phosphorus, sulfur, hydrogen, calcium, potassium, and oxygen—
together with sources and sinks of elements that are present as
minor components in various forms of life.

2. The determination of the rates of organic production and
respiration on land and in sea. How does production on land
change with the climate and with changes in the chemical composition of the atmosphere? What is the relationship between ocean
circulation and organic production in the sea?

3. Biological systems are currently experiencing changes that
are rapid in comparison to evolutionary changes. These changes
represent a perturbation of biological systems, the results of which
may give an important insight into the relationship between biota
and the Earth.

4. Does the increase of nitrogen and sulfur in rain act as fertil-
izer in forests? Will the increased concentration of carbon dioxide
stimulate biotic production? If so, will the carbon-to-nitrogen ratio of plants increase? Will the resulting litter decompose more
slowly, thereby locking up critical nutrient supplies and leading to
a decrease in biotic production? Or will the reverse occur?

The Measurements Required

The external information needed to model these processes includes the major biological sources and sinks of organic carbon and
active nitrogen, and inputs of sulfur and other compounds from
volcanic activity. Urban pollution is a topic all by itself, but is
a major regional source of tropospheric ozone, oxides of nitrogen,
sulfur dioxide, and other ingredients of larger-scale problems like
acid rain. Also required are the transportation and mixing capacity of the atmosphere. Clouds play a key role in catalyzing certain
reactions and in scavenging water-soluble products in precipitation. Although, in principle, the atmospheric transports and cloud
fields are available as part of the modeling and data base of the
physical climate system, in practice, considerable additional effort
is required to make them useful for chemical purposes. Just as
important are the internal measurements that give guidance as to
which chemical processes are most significant and confidence that
they are being modeled correctly. The following measurements are
therefore desirable:

1. Global measurement of changes with time in the minor constituents—both isotopes and elements—of the atmosphere, oceans, and outer-earth layers.

2. A global inventory, as a function of surface slope, of soils of different texture, and water- and nutrient-retaining capacity.

3. Measurements of the quantitative distribution of biomass on the land surface of the Earth.

GRAND THEME 4: INTERACTION OF HUMAN ACTIVITIES WITH THE NATURAL ENVIRONMENT

Global Issues

Human activities since the beginning of the industrial revolution have increased to such an extent that they must now be regarded as important factors in changing the environment. The effects are approaching a significant stage in altering the concentration of ozone and carbon dioxide in the atmosphere, in changing the surface properties by deforestation and erosion, and in other industrial and agricultural activities. Man is a major force now in the chemistry of the atmosphere and in the allocation of resources on land, and increasingly an influence on the ocean. Moreover, the influence can be subtle, as illustrated by the potential vulnerability of stratospheric ozone.

It has become apparent within the last decade that mankind has the ability to alter ozone, and to thus change the level of harmful ultraviolet radiation penetrating to the ground. We can do so by the direct injection of exhaust gases of high-flying aircraft into the stratosphere, by release of chlorinated gases used as aerosol propellants, as industrial solvents, and as working fluids in refrigeration systems, and by complex perturbations to the global nitrogen cycle. These activities lead for the most part to reduction in ozone, but they are offset to some extent by thermal disturbances due to enhanced levels of carbon dioxide, causing a rise in ozone. Assessment of human impact is hampered by lack of understanding of the underlying physical, chemical, and biological influences regulating ozone in the natural state. This matter is critical because the gases responsible for change in ozone—the man-made chlorofluorocarbons and biologically formed nitrous oxide—have lifetimes ranging from 50 to 200 years. The self-cleansing function of the atmosphere proceeds slowly, therefore, and the effects of our actions today will persist for centuries into the future.

Carbon is the largest single waste product of modern society. We have added, by the burning of fossil fuel, over 100 billion tons of carbon to the atmosphere as carbon dioxide since the industrial revolution, with perhaps a quantity of similar magnitude transferred from the biosphere to the atmosphere over this same period as a consequence of land clearance for agriculture. The increase in the burden of atmospheric carbon dioxide is readily detectable. Approximately half of the carbon added to the system remains in the atmosphere and the remainder is presumed to have been taken up by the ocean on its way to the depths of the oceanic abyss, and eventual subduction into the Earth's interior. Attempts to model the process encounter difficulties, however, due in part to deficiencies in our knowledge of the nature of concurrent changes in the global biosphere, interactions with other nutrient cycles—nitrogen, phosphorus, and sulfur, for example—and lack of understanding of the processes of oceanic mixing. The time scales are such as to require a model for the atmosphere, ocean, and biosphere as a coupled system. The matter assumes some urgency since the rising level of carbon dioxide can lead to a change in climate, with associated change in the patterns of rainfall.

The ozone and carbon questions are but two examples of many global issues affecting the environment that must be faced in the years to come. Changes involving soil erosion, loss of soil organic matter, desertification, deforestation, overgrazing, diversion of freshwater resources, and increasing levels of air pollution and acid rain affect the physics, chemistry, and biology of the Earth.

The Measurements Required

Human Impact

Tropical deforestation has recently become a scientific issue of major concern, not only because it significantly decreases biological diversity, and leads to soil erosion and loss of productivity, but also because it is quite possible that it modifies the regional climate in a substantial manner. In addition, the changing global biomass has direct bearing on the carbon cycles and on ocean productivity. An accurate assessment of the rate of change of the forest cover around the globe is therefore becoming an important datum that is needed for a number of disciplines in earth sciences. Ideally, space measurements should be well-suited to document such a change globally, quantitatively, and routinely. However, instruments flown on satellites can only measure radiances either reflected or emitted by the Earth's surface. These measurements have to be rectified for the alterations made by the atmosphere and eventually interpreted in terms of the changes in the biomass or in the properties of the surface cover. It is because of these difficulties that no systematic effort has yet been made to derive quantitative estimates of the rate of deforestation from satellite measurements. The current ground-based estimates range all the way from no "net" change in the biomass to as much as 1 percent per year decrease in forest cover around the world. A narrowing of the range of uncertainty should be the principle objective of any global change monitoring program.

Deserts such as the Sahara may be expanding at a significant rate. There are suggestions that, once the process of desertification starts by the baring of the soil due to human encroachments, the climate becomes drier and the process is self-feeding. The mechanism suggested involves an increase in the albedo of the surface which inhibits convection, thus reducing rainfall. In order to determine whether this mechanism is really at work on a global scale, one needs to measure the change in the surface albedo as a function of time and change in the precipitation distribution around the world. None of these are currently available to authenticate the hypothesis of runaway desertification. Satellite observations integrated with observations on land and the oceans can provide basic data and can monitor soil erosion, desertification, fresh water depletion, variations in the concentration of carbon dioxide and ozone, and the occurrence of acid rain.

Hazards

The larger and larger conurbations that absorb much of the population increase enhance man's vulnerability to natural hazards, such as hurricanes and earthquakes, through dependence on longer and more complex systems of transportation and larger habitational structures. Earthquakes and their associated tsunamis and volcanic eruptions are major hazards to life on this planet. Seismic and geodetic studies have been successful in predicting some volcanic eruptions. It is important to improve these predictions and to apply the techniques globally. It is also important to monitor the effluents from major volcanic eruptions. Predictions can then be made of influences on the global climate. Active volcanoes, instrumented with geodetic devices such as Global Positioning System (GPS) receivers, can be monitored prior to eruption. Our knowledge of earthquakes is much more primitive. Extensive studies of stress, strain, and other observables are required to obtain an understanding of basic mechanisms. Successful prediction remains a goal; however, it is not yet clear whether it will be possible to predict earthquakes with a high degree of reliability. Satellites afford the opportunity to observe geodetic strain changes in the detail essential to improve our understanding. Eventually, this probably will be an important ingredient in any successful prediction program. The space-based geodetic observations must be integrated with a variety of surface observations, including seismic studies.

Tsunamis are often generated by major earthquakes. A global network to monitor and provide tsunami warnings with a high reliability and long lead-time is clearly feasible and desirable. Tsunamis can also be tracked in the open ocean from orbital and ocean floor measurements.

Finally, severe storms constitute yet another major hazard to mankind. Although great progress has been made in predicting and monitoring severe storms such as hurricanes and tornadoes, much remains to be done. Satellite observations, coupled with ground- and ocean-based observations, already provide a much more accurate basis for predicting the occurrence and severity of storms. These studies should also provide the basis for timely warnings of severe flooding.

4
Current and Planned Earth Observing
Satellite Missions: 1986 to 1995

INTRODUCTION

Environmental satellites are the central observing element for a Mission to Planet Earth. They provide the primary means for collecting environmental data on a global, consistent, repetitive, and long-term basis. Prototypes of many of the elements of a Mission to Planet Earth are already proven and operational, or are being planned. This chapter reviews existing space and information systems, as well as those systems that are expected to be operational by 1995. The space-based elements are broken down for the sake of organizational clarity according to six broad areas of inquiry: land, oceans, atmosphere, radiation budget, atmospheric chemistry, and geodynamics.

The discussion that follows is meant to be representative, not comprehensive. All missions and instruments, including those reviewed below and those not discussed, are listed in Table 4.1. A more comprehensive discussion of earth observing missions is provided in the series of NASA reports on the Earth Observing System (EOS).

TABLE 4.1 Observational Programs for Global Data Acquisition: Representative Examples of Approved and Continuing Programs

Program	Agency/Status	Objectives
POES Polar-orbiting Operational Environmental Satellites (e.g. NOAA-7)	NOAA/ Operating	Weather observations
GOES Geostationary Environmental Satellite System	NOAA/ Operating	Weather observations
DMSP Defense Meteorological Satellite Program	U.S. Air Force/ Operating	Weather observations for Department of Defense
METEOSAT Meteorology Satellite	ESA/Operating	Weather observations
GMS Geostationary Meteorology Satellite	NASDA (Japan)/ Operating	Weather observations
METEOR-2 Meteorological Satellite-2	USSR/ Operating	Weather observations
LANDSAT Land Remote Sensing Satellite	EOSAT/ Operating	Vegetation, crop, and land-use inventory
LAGEOS-1 Laser Geodynamics Satellite-1	NASA/ Operating	Geodynamics, gravity field
ERBE Earth Radiation Budget Experiment	NASA-NOAA/ Operating	Earth's radiation losses and gains
GEOSAT Geodesy Satellite	U.S. Navy/ Operating	Geodesy, shape of the geoid, ocean and atmospheric properties
GPS Global Positioning System	U.S. Navy-NOAA-NASA-NSF-USGS/ Completion 1989	Geodesy, crustal deformation
SPOT-1 Système Probatoire d'Observation de la Terre-1	France/ Operating	Land use, Earth resources
IRS Indian Remote Sensing Satellite	Indian/ Operating	Earth resources
Representative Space Shuttle instruments		
ATMOS Atmospheric Trace Molecules Observed by Spectroscopy	NASA/Current	Atmospheric chemical composition
ACR Active Cavity Radiometer	NASA/Current	Solar energy output
SUSIM Solar Ultraviolet Spectral Irradiance Monitor	NASA/Current	Ultraviolet solar observations
SIR Shuttle Imaging Radar	NASA/ Current/in development	Land-surface observations
MAPS Measurement of Air Pollution from Shuttle	NASA/ Current/in development	Tropospheric carbon monoxide
SISEX Shuttle Imaging Spectrometer Experiment	NASA/Planned	Spectral observations of land surfaces
LIDAR Light Detection and Ranging Instrument	NASA/Planned	Surface topography, atmospheric properties

Program	Agency/Status	Objectives
MOS-1 Marine Observation Satellite-1	NASDA (Japan)/ Launch 1987	State of sea surface and atmosphere
LAGEOS-2 Laser Geodynamics Satellite-2	NASA-PSN (Italy)/ Launch 1988	Geodynamics, gravity field
SPOT-2 Système Probatoire d'Observation de la Terre-2	France/ Launch 1988	Earth remote sensing
UARS Upper Atmosphere Research Satellite	NASA/ Launch 1989	Stratospheric chemistry, dynamics, energy balance
ERS-1 Earth Remote Sensing Satellite-1	ESA/Launch 1990	Imaging of oceans, ice fields, land areas
JERS-1 Japan Earth Remote Sensing Satellite-1	NASDA (Japan)/ Launch 1991	Earth resources

Representative International Programs for Measurements in Situ

Program	Organization/ Status	Objective
GEMS Global Environment Monitoring System	UNEP/ Begun 1974	Monitoring of global environment
World Ozone Program	WMO-NASA-UNEP/ Operating	Atmospheric composition
Crustal Dynamics Project	NASA-23 nations/Begun 1979	Tectonic plate movement and deformation
Man and the Biosphere	UNESCO/ Operating	Ecological studies
International Biosphere Reserves	UN/Operating	Long-term ecological studies
ISCCP International Satellite Cloud Climatology Project (World Climate Research Program)	WMO-ICSU/ Begun 1983	Measure interaction of clouds and radiation
ISLSCP International Satellite Land Surface Climatology Project (World Climate Research Program)	WMO-ICSU/ Begun 1985	Measure interactions of land-surface processes with climate
TOGA Tropical Ocean Global Atmosphere Program (World Climate Research Program)	WMO-ICSU/ Begun 1985	Variability of global interannual climate events
GRID Global Resource Information Database	UNEP/ Begun 1985	Information on global resources

TABLE 4.1 (continued) Observational Programs for Global Data Acquisition: Representative Examples of Proposed Future Programs

Program	Agency/ Status	Objectives
TOPEX/POSEIDON: Ocean Topography Experiment	NASA-CNES (France)/Start 1987, Launch 1991	Ocean surface topography
POES: Polar-orbiting Operational Environmental Satellite system — follow-on missions (NOAA K,L,M)	NOAA/ Planned	Advanced capabilities for weather observations
GOES: Geostationary Operational Environmental Satellite system — follow-on missions (e.g., GOES-Next)	NOAA/ Planned	Advanced capabilities for weather observations
RADARSAT — Canadian Radar Satellite	Canada/Start 1986, Launch 1991	Studies of arctic ice, ocean studies, Earth resources
MOS-2: Marine Observation Satellite-2	NASDA (Japan)/ Launch about 1990	Passive and active microwave sensing
GRM: Geopotential Research Mission	NASA/Start 1989, Launch 1992	Measure global geoid and magnetic field
Individual instruments for long-term global observations:		
OCI: Ocean Color Imager	NASA-NOAA/ Planned	Ocean biological productivity
ERB: Earth Radiation Budget instrument	NASA/ Planned	Earth radiation budget on synoptic and planetary scales
Carbon-Monoxide Monitor	NASA/ Planned	Monitor tropospheric carbon monoxide
Total Ozone Monitor	NASA/ Planned	Monitor global ozone
GLRS: Geodynamics Laser Ranging System	NASA/ Planned	Crustal deformations over specific tectonic areas
Laser Ranger	NASA/ Planned	Continental motions
Scanning radar altimeter	NASA/ Planned	Continental topography
Eos: Earth Observing System/Polar-Orbiting Platforms, NASA-NOAA program:	NASA-NOAA/ NASA Start 1989, Launch 1994	Long-term global Earth observations
NASA research payloads	NASA/ Planned	Surface imaging, sounding of lower atmosphere; measurements of surface character and structure; atmospheric measurements; Earth radiation budget; data collection and location of remote measurement devices
NOAA operational payloads	NOAA/ Planned	Weather observations and atmospheric composition; observations of ocean and ice surfaces; land surface imaging; Earth radiation budget; data collection and location of remote measurement devices; detection and location of emergency beacons; monitoring of space environment

Program	Agency/ Status	Objectives
European Polar-Orbiting Platform (Columbus)	ESA/Planned	Long-term comprehensive research, operational, and commercial Earth observations
Rainfall mission	NASA/Start 1991, Launch 1994	Tropical precipitation measurements
MFE: Magnetic Field Explorer	NASA/Start 1993, Launch 1996	Secular variability of Earth's magnetic field
MTE: Mesosphere-Thermosphere Explorer	NASA/Start 1995, Launch 1998	Chemistry and dynamics of upper atmosphere
GGM: Gravity Gradiometer Mission	NASA/Start 1997, Launch 2000	Gradient in Earth's gravitational field

Representative International Programs for Measurements *In Situ*

Program	Organization/ Status	Objectives
WOCE: World Ocean Circulation Experiment (World Climate Research Program)	WMO-ICSU-IOC-NSF-NASA-NOAA/ 1987 enhancement	Detailed understanding of ocean circulation
IGBP: International Geosphere-Biosphere Program (Global Change)	ICSU/ Proposed	Study of global change on timescale of decades to centuries
GOFS: Global Ocean Flux Study	NSF-NOAA-NASA/ Enhancement	Production and fate of biogenic materials in the global ocean
GTCP: Global Tropospheric Chemistry Program	NSF-NASA-NOAA/ Enhancement	Tropospheric chemistry and its links to biota
Ocean Ridge Crest Processes	NSF-USGS-NOAA/ Enhancement	Chemistry and biology of deep-sea thermal vents, plate motions, crustal generation
Sensing of the Solid Earth	NSF-USGS-DoD-NASA/ Enhancement	Large-scale mantle convection, studies of continental lithosphere
Ecosystem Dynamics	NSF/ Enhancement	Studies of long-term ecosystems, biogeochemical cycles
Greenland Sea Project	ISCU/Planned	Atmosphere - sea ice - ocean dynamics

SOURCE: Earth System Science Committee, NASA Advisory Council, *Earth System Science Overview—A Program for Global Change*, NASA, Washington, D.C., pages 34-35, 1986.

CURRENT PROGRAMS

Land Observing Systems

Landsat

The Landsat system is a series of sequentially launched satellites commencing with Landsat-1 in 1972. The onboard instruments of the first spacecraft were a multispectral scanner (MSS) capable of 80-m ground resolution, three return beam vidicons (RBVs), and two wide-band video recorders. The MSS consisted of an electro-opto-mechanical scanner covering a swath 185 km wide in four spectral channels from 0.4 to 1.1 μm. The data were transmitted to earth stations in digital format from a sun-synchronous orbit at an altitude of 918 km. The entire Earth was sequentially covered every 18 days.

Landsat-2 and -3 were identical platforms launched at the same sun-synchronous altitudes, although their instrumentation varied. Landsat-2, launched in 1975, carried a five-channel MSS. The fifth channel was in the infrared range with a 270-m ground resolution; the four other channels, the RBVs, and the tape recorders were identical to those of Landsat-1. Landsat-3, launched in 1978, returned to the four-channel MSS, but used two RBVs operating in a panchromatic mode, increasing the ground resolution to about 40 m. Again, two wide-band recorders were used.

Landsat-4 was a newly designed spacecraft launched in 1982 into a sun-synchronous orbit at 705 km. The instruments included a thematic mapper (TM), which provided a ground resolution of 30 m in seven spectral bands, and a four-channel MSS similar to the one on Landsat-3. Data were transmitted in real time to the ground using a wide-band system in the Ku-band via the TDRSS satellite, or by X-band directly to ground. The TM failed soon after launch in 1982. Landsat-5, with an identical configuration to Landsat-4, was launched in 1984 and is still functioning.

Operational control over the Landsat system was transferred from NOAA to the Earth Observation Satellite Co. (EOSAT) in September 1985. EOSAT is operating the Landsat system on a commercial basis. If they receive the necessary funding, Landsat-6 and -7 are scheduled for early 1990s launches. These spacecraft would have improved capabilities over existing instrumentation and possibly additional sensors, such as a wide-field-of-view imager for ocean phenomena.

Systeme Probatoire d'Observation de la Terre (SPOT)

The Systeme Probatoire d'Observation de la Terre (SPOT), another commercial remote sensing satellite system, is operated by the French space agency (CNES). The first of a series of at least four planned satellites was launched in early 1986 in a sun-synchronous orbit at 832 km. The instruments, two identical pointable multispectral linear-arrays, called high-resolution visible (HRV) sensors, operate in three spectral bands: 0.50 to 0.59 μm, 0.61 to 0.68 μm, and 0.79 to 0.89 μm. The ground resolution is 20 m in a color mode operating across the entire band, and 10 m in a panchromatic mode operating from 0.51- to 0.73-μm bandwidth, and is capable of stereoscopic imaging. Viewing can be forward, backward, and sideways, and each HRV can operate in both modes simultaneously; however, only two sets of data can be acquired at the same time. SPOT images can cover the entire Earth in a 2.5-day repeat cycle. Data are transmitted in real time to 4 ground stations in France, Sweden, and eastern and western Canada, or stored onboard on two wideband tape recorders for non-real-time transmission to the ground. The 4 ground stations are being expanded to 10 in the near future to provide worldwide real time image coverage.

Shuttle Imaging Radar (SIR)

A series of radar imaging experiments called SIR-B, carried out on a Space Shuttle flight in 1985, added another major dimension to NASA's program in earth observations. An earlier experiment, SIR-A, was conducted in November 1981. The data it collected provided the first demonstration that radar sensors can penetrate deep into windblown sand deposits in hyperarid environments. SIR-A imagery of portions of the eastern Sahara Desert revealed the presence of buried drainage channels that provide important clues to the archaeological and geological history of southern Egypt. SIR-B is the first space-borne radar capable of imaging Earth's surface at multiple angles of incidence measured from the local vertical. Initial results were dramatic and showed the SIR-B ability to obtain accurate relief maps of the Earth's surface.

Ocean Observing Systems

The major achievement in ocean observations in the past decade was the 3-month flight of Seasat in 1978, which demonstrated the feasibility of scatterometer measurements for measuring wind and waves, the altimeter for measuring waves and currents, and the synthetic aperture radar for detailed high-resolution measurements of land, ocean, and ice surface. In addition, Nimbus-7, which was also launched in 1978 and is still flying, demonstrated the feasibility of measuring ocean color to be used for estimating sediments and chlorophyll in the near-surface waters. The NOAA operational satellites also have shown the capability for measuring sea surface temperature, which is now a regular operational product.

The untimely demise of Seasat, due to a massive power failure, led to the design of follow-on programs for scatterometers and altimeters. The Geosat altimeter has obtained more than 5 times as much data as Seasat, with comparable accuracy. The NSCAT program, a NASA scatterometer, and the TOPEX/Poseidon mission, which includes a precision altimeter experiment, are discussed below. A follow-on to the Coastal Zone Color Scanner on Nimbus-7 is also being designed, and will fly as part of the EOS.

Atmosphere Observing Systems

The primary observing systems for the atmosphere are the operational weather satellites in polar and geostationary orbit. These are operated by several nations, and the plans detailed here are predicated on their continued operation. As the task group looks to new requirements, it notes that NASA has both the experience and the facilities to deal with the special problems involved in understanding the circulation of the atmosphere; namely, the processing of voluminous data, interpretation of results in meteorological terms, and application of the results to meteorological issues. Investigation and assessment of data from the first Global Atmospheric Research Program experiment are proceeding. A substantial part of NASA's work is devoted to the development of new techniques. For example, an effort is under way to develop and fly an advanced temperature and moisture sounder whose expected performance could approach that of radiosondes, but with far more complete spatial coverage.

The emphasis of research on severe storms and local weather includes meteorological observations from space or high-flying aircraft, and high-technology interactive computer techniques to assimilate and analyze data from multiple sources. One aspect of developing new measurement techniques is the use of aircraft flights for field tests. Also being emphasized are the research applications of the Visible-Infrared Spin-Scan Radiometer Atmospheric Sounder on NOAA's Geostationary Operational Environmental Satellites, and development of new algorithms for determining temperature, moisture, and winds at different heights in the atmosphere for use in numerical models. Flow scales in the atmosphere must be understood if progress is to be made in relating large-scale weather to local weather.

Earth's Radiation Budget

Observations from Nimbus-6 and -7 instruments and from NOAA's operational satellites are the foundation for a continuing series of data sets on Earth's radiation budget that will serve as a resource for climate research. NASA's Earth Radiation Budget Experiments will continue to augment the data sets. Earth's radiation budget also is being addressed in other ways. Evidence from recent Nimbus-7 and Solar Maximum Mission observations confirms that the total output of the Sun varies naturally by several tenths of a percent for periods of up to about 2 weeks. A number of instruments are being designed to monitor the long-term trend of solar variation and to determine its effect on climate systems, and these are discussed later in the chapter. Research programs have been initiated to develop an understanding of and models for the processes by which clouds are formed and interact with incident or reflected radiation, and to study the sources, compositions, and radiative effects of aerosols that volcanic explosions inject into the stratosphere. In addition, the International Satellite Cloud Climatology Project is expected to develop a global cloud climatology data set.

Atmospheric Chemistry

Investigators are developing techniques for measuring major trace species in the troposphere. Field measurements to test the most promising instruments will be followed by a 6-year program of measurements by aircraft to characterize the chemistry of the troposphere on a global scale. Research on the stratosphere and mesosphere also continues and has increasingly used more realistic two- and three-dimensional models. The chemical, radiative, and dynamic computer codes used in those models are being improved continually, with the goal of developing fully coupled chemical, radiative, and dynamic three-dimensional models that simulate the atmosphere very precisely. Also NASA, in cooperation with European, Canadian, and Japanese investigators, is using a variety of instrument techniques on balloon, rocket, and aircraft flights to obtain measurements of trace species in the stratosphere that will allow accurate comparisons with current experimental techniques.

Data from Nimbus-4, -6, and -7, and the Stratospheric Aerosol and Gas Experiment have been validated and are becoming available for detailed analysis. Solar Mesosphere Explorer data on ozone, nitric oxide, and water vapor will also soon be available for analysis. In addition, two instruments, the Imaging Spectrometer Observatory and the Atmospheric Trace Molecule Spectroscopy experiment, have been developed for use on the Shuttle to measure those species in the mesosphere and stratosphere. The Observatory already has flown on Spacelab 1, and the spectroscopy experiment may fly on a future Spacelab mission.

Geodynamics

Laser ranging, lunar ranging, and microwave interferometry (VLBI) are being used to measure the motions of Earth's polar axis, variations in the length of day, and the motion and deformation of Earth's crustal layer. A worldwide network of over 20 cooperating space agencies participates in NASA's global geodynamics research. A second Laser Geodynamics Satellite (LAGEOS), being built by Italy, is expected to be launched in 1993. Data from laser tracking of satellites, and altimeter data from Seasat and the third Geodynamic Experimental Ocean Satellite (GEOS-3) improved the accuracy of models for global gravity fields used in studies of earth and ocean processes. Similar data acquired by the Magnetic Field Satellite (Magsat) were used in studying secular and temporal variations of Earth's main field and inhomogeneities in Earth's crust.

POTENTIAL INITIATIVES: 1986 TO 1995

The initiatives below are those that have been developed from the knowledge base gained from the programs described above. The first two of these, the Upper Atmosphere Research Satellite and the scatterometer, were approved as new starts in 1985. NASA had no new starts in 1986, but did obtain a new start for TOPEX/Poseidon in 1987. The other initiatives are further down the queue, but are expected to be strong candidates for new starts in the period 1988 to 1995. Finally, the task group anticipates that the Earth Observing System will be the major new start of this group. It is discussed in Chapter 5.

Upper Atmosphere Research Satellite (UARS)

This program's goal is to extend scientific understanding of the chemical and physical processes occurring in Earth's stratosphere, mesosphere, and lower thermosphere. Its primary objective is to observe the mechanisms that control the structure and variability of the upper atmosphere, the response of the upper atmosphere to natural and human-related perturbations, and the role of the upper atmosphere in climate and its variability. It will use remote sensing instruments currently in development, including two instruments being provided by British and French investigators, to measure trace molecule species, temperature, winds, and radiative energy input from and losses to the upper atmosphere. It also will make in situ measurements to determine magnetospheric energy inputs to the upper atmosphere. Plans include extensive interaction among experimental and theoretical investigations, and an interactive central data facility with direct on-line access via remote terminals to facilitate that interaction among investigators. It is expected to fly in the early 1990s.

Scatterometer

Upper ocean currents, as well as surface waves, are generated by the stress that winds exert on ocean surfaces. As earlier instruments aboard aircraft and Seasat have shown, a scatterometer can measure the small-scale roughness of a sea surface; the associated wind velocity, or stress, then can be calculated. Modern oceanographic measurements show that ocean currents are much more variable than they previously were thought to be. An ability to obtain wind velocities will permit calculation of the velocities of the time-dependent, wind-driven, upper ocean currents. Knowledge of those velocities will substantially improve understanding of the momentum coupling of the atmosphere and oceans. Knowledge of wind velocities also will improve forecasts of such factors as wave conditions and the intensity and location of storms. Scatterometer data would provide a unique global perspective of the oceans, significantly improving understanding of how the oceans work. A scatterometer is tentatively planned to be flown on the U.S. Navy's Remote Ocean Sensing Satellite (NROSS). Other plans include flight of a scatterometer aboard the European Space Agency's (ESA) ERS-1 satellite. Both the NROSS and the ERS-1 are expected to be launched in the early 1990s.

Ocean Topography Experiment for Ocean Circulation (TOPEX)/Poseidon

The Ocean Topography Experiment, a joint U.S./French initiative, is expected to provide significant capabilities for observing the circulation of the oceans on a global basis. Its objectives will be to measure ocean surface topography over entire ocean basins for several years, integrate those measurements with subsurface measurements, and use the results in models of the oceans' density fields to determine the oceans' general circulation and variability. The information from all those activities will be used to develop an understanding of the nature of ocean dynamics, calculate the heat transported by the oceans, understand the interaction of currents with waves, and test the capabilities available for predicting ocean circulation. TOPEX/Poseidon is planned to be launched on Ariane in 1991.

Ocean Color Imager

The success of the Coastal Zone Color Scanner, which was launched on Nimbus-7 in 1978 and now is in its eighth year of operation, clearly indicates that a follow-on instrument could determine global primary productivity, which forms the base for the various marine food chains. The synoptic, global measurements of chlorophyll concentration that a satellite color scanner can provide will serve as the primary data base to which complementary ship, airplane, and buoy data can be added to yield primary productivity estimates of high accuracy for key oceanic regions.

An improved version of the Coastal Zone Color Scanner, the Ocean Color Imager, has been designed. Plans are being formulated to make it possible, for the first time, to relate wind forcing data acquired by a NASA scatterometer to data on ocean current response from the planned TOPEX/Poseidon mission, the redistribution of oceanic nutrients by the currents, and the resulting changes in primary productivity from the Ocean Color Imager. With appropriate in situ observation, it will be possible to quantitatively relate biological variability to the physical characteristics of the global oceans.

Shuttle-Spacelab Payloads

Basic processes in which electromagnetic energy and particle beams interact with plasmas occur in many systems within the universe, but can be studied most easily in the most accessible space plasma—that near Earth. Spacelab's capabilities are well suited for making those studies. A beginning was made with the flight of the OSS-1 pallet, which used a small electron gun to study vehicle charging and wave generation. Spacelab 1 had a Japanese electron accelerator with pallet-mounted diagnostics, and a future Spacelab may include an electron gun and a plasma diagnostic package on a subsatellite. Under current planning is a more ambitious mission, called the Space Plasma Laboratory, on which those instruments will be joined by including a VLF-HF wave injection facility being developed in cooperation with Canada. Because of Spacelab's versatility, the mix of instruments can be changed between flights and the entire payload can be upgraded in an evolutionary fashion. Also planned is the assembly, into a single payload, of several solar radiance instruments (the French-developed Solar Ultraviolet Spectral Irradiance Monitor, the Active Cavity Radiometer, and the Belgian-developed Solar Constant Variation instrument) and two atmospheric instruments (the Atmospheric Trace Molecule Spectroscopy experiment and Imaging Spectrometer Observatory).

C-2

Tethered Satellite System

The Tethered Satellite project is a cooperative undertaking between the United States and Italy to provide a new facility for conducting earth science and applications experiments. The Tethered Satellite will make measurements as far as 100 km from the Space Shuttle. It will make possible long-term scientific experimentation not heretofore feasible. This will include the generation and study of large-amplitude hydromagnetic waves, magnetic-field-aligned currents, and high-power, very low frequency and extremely low frequency waves in the ionosphere-magnetosphere system. It also will permit studies of magnetospheric-ionospheric-thermospheric coupling and atmospheric processes below 180 km; high-resolution crustal geomagnetic phenomena; and the generation of power using a conducting tether. Italy has agreed to provide the satellite for the planned atmospheric (tethered downward) and space plasma (tethered upward) missions.

Magnetic Field Satellite

The first Magnetic Field Satellite, Magsat-1, acquired the initial detailed, global data on the scalar and vector magnitudes of Earth's magnetic field. However, that field undergoes major changes over the period of a few years due to variations in the motions of the outer core. The position of the magnetic pole drifts westward, but the rate of drift is not constant. Resulting uncertainties in magnetic maps limit their usefulness to from 3 to 5 years. However, those changes provide information on important and enigmatic properties of Earth, such as the origin of the main magnetic field and its variations with time; the structure and electrical properties of the mantle; and the relationship among variations in the magnetic field, the mass distribution of the atmosphere, and the rotation rate. The Magnetic Field Explorer will obtain scalar and vector field data that, in conjunction with data from Magsat-1 and the Geopotential Research Mission, will be used to examine magnetic field changes for periods ranging from months to decades. It also will provide an updated data set required for a future magnetic field survey.

Geopotential Research Mission (GRM)

Accurate knowledge of Earth's gravity and magnetic fields is essential to scientific studies of the planet, particularly those involving the solid Earth, the oceans, and energy and mineral resources. Earth's gravity field is known to an accuracy of 5 to 8 mgal for resolutions of 500 to 800 km, and the geoid (mean ocean sea level) to an accuracy of about 50 cm. Those accuracies are inadequate to resolve key scientific questions relating to the motion of Earth's crust (mantle convection) and the structure and composition of Earth's interior. Magsat-1 provided a map of crustal magnetic anomalies that showed a high degree of correlation with large-scale geological and tectonic features. However, its orbital altitude was too high to yield a map with the accuracy and resolution required for both solid earth science and geological prospecting. Greater accuracy and resolution are needed, and they can be achieved only by a mission at a significantly lower altitude.

The Geopotential Research Mission will provide the most accurate models yet available of the global gravity field, geoid, and crustal magnetic anomalies. It will employ two spacecraft approximately 300 km apart in the same 160-km circular polar orbit. To determine the gravity field, a drag-free sphere will be positioned at the center of mass of each spacecraft in a cavity that will shield it from all surface forces and therefore permit it to be affected only by gravitational forces. The relative motion of the spheres as they are accelerated and decelerated while passing over a gravity anomaly will be a measure of the size and intensity of the anomaly. The accuracy to which the position of each sphere in the along-track direction can be measured by Doppler tracking will be 1 μm/s every 4 s. That accuracy in the Doppler data will permit analysis to determine the global gravity field to approximately 1 mgal and the geoid to approximately 5 cm, both to a resolution of 100 km. Earth's magnetic field will be surveyed by scalar and vector magnetometers, similar to those flown on Magsat, mounted at the end of a rigid boom extending from the leading spacecraft. The magnetic field data will have an accuracy of 2 nT and a resolution of 100 km.

Earth Observing System (EOS)

The Earth Observing System is an integrated set of experiments that builds on all of the above to form the basis of the Mission to Planet Earth. It is described in the following chapter.

COMPUTERS, COMMUNICATIONS, AND DATA MANAGEMENT

The Space Science Board's Committee on Data Management and Computation (CODMAC) recently completed two extensive studies of space data issues. These reports, *Data Management and Computation—Volume 1: Issues and Recommendations* (1982) and *Issues and Recommendations Associated with Distributed Computation and Data Management Systems for the Space Sciences* (1986), address data issues over a broad range of space science disciplines. They conclude that data management problems account for many of the shortcomings in the science returns of space observation programs. The Task Group on Earth Sciences concurs with the CODMAC findings, and notes that the high data rates from earth observing satellites will strain the system more than any other discipline.

5
Elements of the Mission to Planet Earth

SYSTEMS DEFINITION

The grand themes can be addressed adequately only by a Mission to Planet Earth that includes an integrated and interrelated set of satellite and surface observations. These will involve polar and geosynchronous orbiters, special purpose orbiters, data relay satellites, ocean drifting instruments, pop-up buoys, tethered buoys, ocean bottom instruments, optical fiber links to islands and buoys (up to 1000 km seems to be feasible), automated ground stations, and simple ground stations such as corner reflectors, rain gages, and tide gages. Microchip, computer, and low-power technology can make these sea and land observations extremely powerful in their capabilities, and satellite relay links will make it possible to collect data from a large area of the Earth's surface. In addition, modules can be developed for ships and aircraft that can be linked into the global data network.

The measurement strategy for the implementation of an observing system to address the grand themes involves a system that is composed of about 8 orbiters, 1000 ocean systems, 1000 automated land stations, thousands of simple surface stations and a variable number of itinerants (ships, planes, balloons). The elements of the entire system are summarized in Table 5.1, including

TABLE 5.1 Elements of a Mission to Planet Earth

Quantity	Element
1	Space Station
1-2	Special-purpose satellite missions
5	Geosynchronous orbiters
2-6	Polar orbiters
18-24	GPS constellation
2-3	Tracking and data relay satellites
1,000	Floating buoys, "pop-ups"
100	Moored buoys
100	Ocean bottom stations
100-1,000	Smart ground stations
1,000-10,000	Simple ground installations
	Aircraft
	Balloons
	Rockets
	Ships, research
	Ships, opportunity

the various components of PLATO. The association of a variety of measurements with satellites and in situ stations is summarized in Table 5.2. Also included in the latter table are the types of instrumentation required to obtain the measurements.

Several of the subsystems and experiment packages are well known and described in NASA documents. Others such as ocean drifting instruments, pop-up buoys, and smart ground stations are relatively new concepts. Some of these would be called landers, penetrators, unmanned stations, or rovers if they were being developed for other planets. A part of the strategy for such an interlocked system is given in the NASA Earth Observing System (EOS) report, but the subsystems described therein are just a portion of the whole earth experiment the task group is proposing.

As has been repeatedly emphasized, it is important that these measurements be globally complete, simultaneous, and continuous. The next task is to suggest what sort of observing systems might satisfy these requirements. The full specification of such an implementation in a period as remote as 1995 to 2015 would take an expertise in instrumentation and space systems not present in this task group. The task group can, however, use reports applicable to the shorter time range of the 1990s and make some reasonable conjectures about further advances in measurement and

TABLE 5.2 Summary of Instruments and Measurements for the Mission to Planet Earth

Measurement	Measurement Systems					
	Polar Orbiter	Geosyn-chronous	Other Satellites	Aircraft Balloons	Ocean Stations	Ground Stations
Magnetic field	G		G	G	G	G
Gravity field (geoid)	G	G, A	G	G	G, A	G, A
Stratospheric chem.	C, P			C	P	P
Aerosols	M, P	M		T	T	T
Winds	A, P		A	T	T	T
Severe storms	R, M	R, M	D	T	T	T
Clouds	M, P			T	T	T
Precipitation	R, M, P	R, M	M, D		T, S	T
Particulate matter	M, P	M		T	T	T
Tropospheric chem.	C, P			C	C	C
Ocean currents	M, A	M, A	A, D		A	
Ocean chlorophyll	M	M		O	O	
Ocean salinity	M	M			O	
Lake levels			A			O
Sediments	M	M	D		O	
Sea state	R	R	D		O	
Sea ice	R, A, M	R, A, M	A		O	
Glaciers	R, M, A		A			A
Snow	R, M	R, M				A
Topography	A		A			A
Surface temperature	M	M		O	O	T
Albedo	M	M				T
Surface geochemistry	M	M				
Geological features	R, M	R, M				
Cultural features	R, M	R, M				
Vegetation	M	M				T, C
Soil moisture	M	M				
Soil erosion	M	M				
Surface strain	A		A, D			A
Seismic wave velocities		D		S	S	
Tectonic deformation			D, L		S	S

NOTE: Instrument categories are as follows:

M	Multispectral imaging	T	Meteorological instruments
R	Radar imaging	O	Oceanography instruments
A	Altimetry ranging	S	Seismographs-acoustic detectors
P	Vertical profile remote sensing	G	Gravimeter-magnetometer
D	Data links	C	Chemical composition instruments
		L	Locations, precise geodetic

data processing capabilities, and then suggest the broad outlines of a Mission to Planet Earth after the turn of the century.

The two reports on which this chapter is partially based are *Earth Observing System* (NASA Goddard Space Flight Center, Tech. Memo. 86129, August 1984 (2 volumes)); and *Earth Systems Science Committee Working Group on Imaging and Tropospheric*

Sounding Final Report (Caltech Jet Propulsion Laboratory D-2415, January 1985). This second report is, to an appreciable extent, a critique of the first.

ELEMENTS OF THE SYSTEM

Satellite "Earth Observing System" (EOS)

The data base that has been built by operational land and meteorological satellite systems since the early 1970s must be preserved and continued. These data should be placed and maintained in forms that optimize their use in combination with the data generated by the missions already specified for project starts in the period 1986 to 1995, and with the EOS information system described below, to be implemented in the mid-1990s.

Earth observables can be characterized in two broad categories: *quasi-static*—properties varying slowly enough that mappings at intervals of several years suffice (e.g., rock types, vegetation regimes, magnetic field); and *dynamic*—properties varying on time scales of minutes to seasons (e.g., precipitation, vegetation state, snow cover). EOS is primarily directed to quantities in the second category. The required resolutions and instrument sensitivities anticipated in the 1990s are such that the most efficient orbit appears to be sun-synchronous at altitudes of 600 to 1000 km: i.e., inclination about 95°, circular, about 14 orbits per day. Such an orbit sees the atmosphere and surface below it always at the same local time, which greatly eases data analysis; 2:00 p.m. is recommended by the EOS report. Some parameters can be observed from geosynchronous altitudes at 36,000 km. An equatorial geosynchronous orbiter can observe effectively to latitude 60°, and a set of five such orbiters (e.g., the current operational weather system) gives good global coverage within these latitude bands.

Principles important to any observing system are global synopticity, nested coverage (coordination of lower and higher resolution systems), quality control—most importantly, calibration by in situ measurements—and integration with data interpretation and data continuity. The instruments given in Appendix A are currently recommended for EOS. The altitude of 600 to 1000 km is based primarily on the swath width appropriate for the MODIS, HMMR, and LASA (see Appendix A) instruments. All of this instrumentation would not be carried on the same polar-orbiting sun-synchronous spacecraft. For example, the radar altimeter will be carried on the non-sun-synchronous TOPEX satellite. The report does not address the needs for observations from geosynchronous orbits, specialized orbiters, or nonorbiting devices, all of which are important aspects of a total earth observing system.

As this report and many others have emphasized, the global nature of atmospheric, oceanic, and land processes makes it necessary to view Earth as a single interactive system in order to describe, understand, and predict significant trends in its state. At the same time, the complexity of the internal processes occurring in each component of the system makes it prudent to study these processes separately. We therefore should pursue a strategy of assessing the capabilities of the entire spectrum of remote sensing techniques, identifying the directions in which real progress can be made. The ultimate goal is to develop an observational system that can relatively quickly provide an improved state of knowledge about planet Earth and its main components. This leads to the following recommendations:

1. A major effort needs to be devoted to amalgamation of remote sensing data from several spectral regions and from more than one source in order to improve the accuracy of derived parameters. In addition, space and in situ measurements must be combined to provide optimum measurement systems for obtaining the basic data.

2. In view of the long-term efforts that used to be devoted to the derivation and utilization of remote sensing data, there is a critical need to develop acceptable calibration methods to guarantee long-term stability of the measured data and of the retrieved geophysical parameters. In particular, we need to establish an acceptable approach for providing "relative calibration" of the solar channels on Landsat, AVHRR, and HIRS.

3. Space and in situ conventional data must be related in order to understand the physical processes producing remotely sensed signals and to provide the instrument verification and calibration needed to maintain the integrity of the data sets.

4. The requirements for the radiation budget at the top of the atmosphere, surface flux, sea surface salinity, and surface pressure should be added to the set of important parameters. In addition, there is a strong sense of urgency regarding the need for rapid development and implementation of observational tools for soil

moisture, precipitation, land surface temperature, and spectral emissivity.

5. EOS should add high-spectral-resolution visible and infrared instruments for accurate determination of important surface and atmospheric parameters. Active lidar systems should also be advanced and tested because of the potentially unique contributions they offer.

In general, the task group notes that we must design future remote sensing instruments as integrated systems with complementary sensor packages. This includes coordinated data analysis algorithms, calibration, and validation methods that will permit us to observe the planet Earth in "all its dimensions." Because the requirements for earth science research are so demanding in terms of accuracy and precision, and are of such long-term nature, we cannot rely on placing together the needed data from fragmented sets of observations.

The EOS space-based infrastructure should be designed to expedite deployment of new instruments using onboard data processing and data compression, with overlapping data from consecutive sensors. EOS should also consider the deployment of instruments that include, as an essential part of the system, elements deployed on the surface. As one example, the task group recommends the development and deployment of a system of ocean stations to collect surface and subsurface information.

In order to accomplish the Mission to Planet Earth we will need development of new technology to measure some of the more elusive parameters, combine data from the ground network of stations, buoys, and pop-ups with those coming from satellites, and develop further the data processing, merging, selection, and distribution capabilities. For instance, satellite data collection from autonomous drifting platforms at sea has become routine in the past decade through the French System ARGOS, carried by the TIROS/NOAA A-J polar-orbiting satellites. Fewer than 1000 platforms are now monitored in this way. As new global programs aimed at understanding climate and global change become fully operational, the task group expects to see an increase of a factor of 10 in data rate. Moreover, the collection of oceanographic and seismic data by sea floor and island stations that use fiber optics and other high-data-rate collection techniques is expected to place a demand on the data system at least equal to that of the drifting platforms. Thus the task group expects that the 1990s will see an urgent need for an increase of at least one order of magnitude in global satellite data collection and transmission capabilities. *The task group recommends that planning begin now for a follow-on to ARGOS that will be capable of handling the data from the approximately 10^4 platforms, and the high (10 times present rates) data rate platforms that are expected in the mid-1990s.*

Smart Ground Stations

A smart ground station (SGS) would have a modular design and an on-site computer capable of a variety of "intelligent" decisions. For example, a microcomputer based on a 68020 Motorola chip would match the computer power of a VAX/780. Filtration, averaging, some quality control, and scheduling of individual measurements could take place on site. These stations could also have a built-in recording system, such as a large-capacity cassette or magnetic tape unit, but the primary mode of data transmission would be satellite telemetry. It is anticipated that there would be 100 to 1000 SGS installations worldwide.

With the computing power described above, the SGS could support many simultaneous measurements, including seismic, geodetic, meteorological, hydrological, and soil properties. These are examined below.

Seismic Measurements

An SGS could measure the structure of the Earth's interior, earthquake source mechanisms, strong ground motion in seismic areas, volcanic activities, and tidal forces. A three-component, broad-band seismic station would be capable of resolving 10^{-9} m/s^2 at a period of 25 s. Strong motion instruments should also have broad-band capabilities and remain on scale for accelerations above 1 g. Each channel would be sampled continuously at a rate of 20 samples per second using up to 3 bytes per sample, although a reduction through data compression might be feasible. Strong motion instruments would be triggered if acceleration were to exceed 0.01 g. The sampling rate would be 200 samples/s and the burst data rate would be 1800 bytes/s. Data could be stored in an on-site buffer and transmitted at a lower rate.

In volcanically active areas, an SGS could monitor a local

network of short-period seismometers. The network could have continuous in situ recording with the telemetry activated upon detection of an eruption.

Geodetic Measurements

Stations making geodetic measurements could monitor ground motions, strains, uplift, volcanic activities, fault zones, plate tectonics, and post-seismic and post-glacial rebound. Instrumentation would be dependent on tectonic setting, but it can be expected that emphasis would be on distances comparable to fault lengths and depths of minimally significant earthquakes. For example, investment will be in position-difference instruments. Strainmeters and tiltmeters would be deployed in tectonically active regions, with sensitivity and type depending on their proximity to the fault zone. A typical sampling rate might be one sample per 10 s. Depending on the anticipated level of tectonic deformation, a GPS receiver could be permanently deployed, in which case several measurements per day could be performed, or it could be brought to the site at monthly to yearly intervals. In any case, the on-site computer could perform all the necessary signal processing at an expected accuracy of 1 to 3 cm. Electro-optical ranging devices may nonetheless continue to be more precise than the GPS at distances less than 10 km. However, mobile GPS receivers will be needed to obtain spatial resolution complementary to the temporal resolution of the fixed instruments.

Meteorological and Hydrological Measurements

The purpose of obtaining long-term calibrated data in these areas is to understand the coupling of radiative, dynamical, and chemical processes in the atmosphere; to improve the accuracy and extend the range of weather forecasting; to assess the influences of changes in sea surface temperature, ocean surface currents, and sea and land ice cover on climate; and to determine what factors control the hydrological cycle. The SGSs will be indispensable in obtaining the 10 to 50 years of continuous observations that are needed to collect the requisite amount of data.

Atmospheric measurements are necessary for the following parameters: temperature and humidity profiles, surface winds, aerosols, minor constituents, long-wave and short-wave radiation, and the amount, height, emissivity, albedo, and water content of clouds. With regard to hydrology, measurements must be made of ice and snow extent, thickness and dynamics, precipitation, evapotranspiration, water runoff, and ocean and land surface temperatures and albedos. The reader should consult *A Strategy for Earth Science from Space in the 1980's and 1990's, Part II: Atmosphere and Interactions with the Solid Earth, Oceans, and Biota* (National Academy Press, 1985) for the recommended measurement requirements for each of these parameters.

Measurement of Soil Properties

Smart ground stations will help determine the relationship between climate, vegetation, soil moisture, and topography. They will also assist in our understanding of the effects of changes in land surface evaporation, albedo, and roughness on local and global climate. Long-term data sets for the following soil-related parameters will be required: soil types and areal extent, moisture of the surface and root zones, texture, color, elemental storage, temperatures, infrared emissivity, and albedo. The recommended measurement requirements are set forth in the National Academy Press publication cited above.

Simple Ground Installations

In addition to those measurements requiring the high-data-rate facilities of the smart ground stations, simpler, more traditional measurements at a large number of sites will still be necessary. There may be as many as 10,000 of these installations, as suggested in Table 5.1. One example is given here.

Magnetic Field

The magnetic field has been monitored at several dozen sites around the world for more than a century. However, this distribution is very nonuniform, and consequently inferences of lateral heterogeneities in the rate-of-change in the magnetic field have been quite uncertain. This problem would be appreciably relieved by a magnetic monitoring satellite, a relatively high altitude (up to 1000 km) spacecraft with a lifetime of decades. Because of temporal fluctuations generated by ionospheric currents, however, there will still exist aliasing of internally generated changes inferred from a single orbiter. Hence if, as proposed here, there is established a global set of ground stations, it would be desirable to include at least 100 magnetometers as uniformly spread as feasible (i.e., at 2000-km intervals). This modest investment would greatly enhance the value of a magnetic monitoring satellite.

Ocean Bottom Stations

In this version of the ocean bottom geophysical observatory, the signals would be transmitted by a fiber optic cable to the nearest island, from which the data would be transmitted by satellite. Whether the station processor is deployed on the ocean bottom or on the island will depend on the power requirements. Generally, the capacity of the processor should be similar to that of an SGS, except that if power consumption considerations should so indicate, part of the necessary computations could be performed at the data collection center.

A primary purpose of this type of observatory would be to monitor seismic activity on the ocean floor, including the structure of earthquake source mechanisms, strong ground motions in seismic areas, volcanic activities, and tidal motions. The seismic stations would have three-component, broad-based instrumentation capable of resolving 10^{-9} m/s^2 acceleration at a period of 25 s. Experimental measurements indicate that seismic noise in a deep ocean borehole is comparable to the noise level at quiet sites on land. There is therefore no reason to lower the standards established for the land observations. Strong-motion instruments deployed in seismically active areas should also be broad-band type and remain on scale for acceleration above 1 g. Data rates should be identical to those specified for SGS.

Additional uses of these stations could be to measure a variety of ocean bottom characteristics such as pressure, temperature, and chemistry. The data rate for these could be 1 sample per 10 s, which would pose a negligible contribution to the data flow.

Moored Buoys

Moored buoys provide the means for measuring properties of the ocean at a fixed point. They can be combined with the ocean bottom stations described above. As with the stations, and depending on power requirements and proximity to land, there may be direct relay of the data by satellite, or a link to the nearest land mass may be established by fiber optic cables.

The buoys could be used primarily for oceanographic purposes, as they are at present. They could characterize processes below the sampling depths of satellites, vertical mixing rates of the surface layer, exchanges between the surface and deep water, and mesoscale features. Surface measurements would continue to be used for "ground truth" verification of satellite data. Seismic measurements, similar to those of the ocean bottom package, could be performed as well.

Floating Buoys

These devices, in addition to the oceanographic capabilities described in connection with moored buoys, allow for mapping of ocean currents. Because their position changes with time, they must be able to be easily located. This can be accomplished with an appropriate space-based system such as ARGOS. Drifting buoys can be at the surface, in which case they measure surface air and sea parameters as well as the ocean currents, or they can float at a given density level below the surface, monitoring temperature and currents. The subsurface buoys are programmed to come to the surface periodically to report their position and data they have collected previously—hence they are called "pop-up" buoys. While a full seismic package cannot yet be deployed, a hydrophone could provide information useful in locating earthquakes. Technical details still need to be developed.

Ships: Research and Voluntary Observing

There are about 300 oceanographic and marine geophysics research ships operating today, and there are many others that are willing to take measurement on a not-to-interfere basis (voluntary observing ships). They are usually on special research expeditions that collect data and that may or may not be linked into the satellite data system. These can be considered additional sources of observations that can provide such basic data as temperature, salinity, precipitation, winds, wave data, and sea state. Modules can be designed for shipboard use that automatically collect and transmit these data, or they can be treated as manned observatories.

Global Geodetic Observations at the Centimeter Level

The Global Positioning System (GPS) that is being deployed by the DOD provides tremendous opportunities for accurate geodetic measurements on a worldwide basis. For ordinary surveying purposes, inexpensive instruments will greatly reduce costs after 1988. For the geophysical purposes of intent here, the limiting factor is currently tropospheric resolution. However, by 1995 it can be expected that water vapor radiometers, or substitutes for them, will be sufficiently inexpensive to enable widespread campaigns with the GPS system. For the primary purpose of providing data on geodetic displacements in areas of tectonic activity where most earthquakes occur, a dense network of control points at intervals of 1 to 30 km (dependent on site) is indispensable if the unraveling of complicated strain rate patterns is to be feasible. Accuracies of a few parts in 10^8 (e.g., a function of 1 mm over 10 km) should be achieved by the turn of the century. The GPS system can also provide data on inflation of volcanoes in order to instrument the volcanic eruption hazard on a worldwide basis. Secondary applications on land include measurement of reclining tectonic plate motions, mass loss of glaciers and ice caps, and post-glacial rebound. Observations on the oceans and their boundaries can provide data on global changes in sea level associated with changes in ice volume or other causes, ocean tides, and ocean currents.

There is a need for permanent stations, including moored buoys, that monitor geodetic positions continuously and also for mobile stations that can monitor position on closely spaced networks in areas of concern, such as the San Andreas fault in California. In addition, the GPS system will provide accurate locations for satellites, aircraft, and ships, and thereby greatly improve their measurement capabilities.

The development of precise GPS receivers is revolutionizing the field of geodesy. A large number of high-accuracy measurements of vector position and baseline length can be made economically with such receivers. Given a sufficient number of constraining measurements, major progress in understanding local and regional crustal deformation and perhaps plate driving forces should be possible.

Receiver performance and uncertainty in the orbit ephemeris currently limit GPS accuracy, but both will improve significantly over the next several years. The limiting error source in the near future is thus expected to be uncertainties in signal delay related to propagation through a variable atmosphere. Improvements in water vapor radiometers (WVR) and improved atmospheric models are required before ultra-high-precision GPS geodesy ("super GPS") becomes a reality. It may well be that a series of GPS and WVR observations stretching over several years in a variety of climatic zones will be required to separate systematic errors from plate motion and to fully understand atmospheric effects on GPS location techniques. Continuous observations would be necessary in order to characterize the dynamic effects of weather with periods of several days, in addition to longer period seasonal atmospheric effects. Permanent GPS facilities at key sites would therefore be required.

The geodetic application of GPS to obtain differences of position interferometrically is a by-product use not anticipated by the DOD. Eventually, it can be expected that a system with an optimized signal and a spacecraft easier to model for radiation pressure will become desirable for geodetic purposes.

For certain areas, such as the San Andreas fault region, it may be useful to have some GPS receivers operating frequently or continuously at fixed points in an automatic or semiautomatic mode, similar to tide gages now. This mode would reduce errors due to local disturbance of monuments and imprecision of site reoccupation. But the major application of GPS to geodynamics will be by mobile systems, reoccupying a network of sites at intervals of months to years. For this mode, it may be that a more economical system of comparable accuracy will be laser ranging from spacecraft to retroreflectors on the ground. This technique should be developed, and then the costs should be compared with those for the GPS mode.

It has now been confirmed from analysis of satellite ranging involving several tectonic plates that relative motions over an 8-year time base agree within 1 cm/yr, with plate motion rates based on sea floor magnetic lineations an a 10^5-year time scale. It can be expected that this picture will be greatly refined by 1995. Hence, effort in the period 1995 to 2015 will be much more focused on regions of known deformation in the Holocene era.

A more speculative aspect of geodetic measurements for geodynamics research is the ability to monitor the position of sea floor geodetic points. This could be done acoustically using three or more precision transponders mounted permanently to the sea floor, and a central surface platform such as a buoy for transponder interrogation and travel time measurement. The surface platform must be positioned relative to the nearest land via GPS, so that in effect the platform would act as an air-water transfer point for positioning. For several possible deployment platforms, particularly smaller, more economical buoys, the GPS unit would need to operate untended for the duration of the experiment. The ability to operate for longer periods in an untended mode would reduce the number of expensive ship visits for data transfer. The major limitation on the operational life of such an experiment would then be the lifetime of battery packs for the sea floor transponders, currently 2 to 3 years for such low duty cycle applications.

Thus, a number of considerations suggest that a key part of NASA's program in solid earth geophysics should include development of a GPS-based geodetic system capable of operating at remote sites for long periods (months to years) in an untended, fully automated mode. The benefits that would accrue from such a development include improved atmospheric models and accuracy, reduced cost of field operations, a better description of the temporal spectrum of plate motion, and the ability to carry out precise location of surface buoys in support of a sea floor geodetic program. Land GPS stations could be deployed advantageously with the global digital seismic array as well as with many other types of instrumentation because they can share a common data transfer system.

While GPS measurements will become more focused on tectonic problems that are more precisely defined by a context of geologic and seismic surveys, a need for monitoring of global-scale motions will remain. For this purpose, VLBI is becoming preeminent because it has all-weather capability and a multiplicity of sources. Lunar laser ranging will continue to be of value to monitor systematic error sources and study the Moon's orbit. VLBI is also able to provide orientation and precise differences of location to timing and factors determining GPS orbits.

The PLATO System

Once in place, the Permanent Large Array of Terrestrial Observatories (PLATO) system can be viewed as a gigantic terrestrial observatory, which, in addition to its normal functions, can be treated as a telescope or accelerator. It can be reconfigured for special preplanned experiments and preprogrammed for "targets of opportunity," such as volcanic eruptions or tsunami tracking, that require, for example, higher data rates for certain periods of time. The system and potential experiments can be upgraded as technology improves.

Many of the technology needs for PLATO have been identified in previous sections and in other studies, particularly the NASA report, *Earth Observing Systems*. To summarize, these needs will require developments in two broad areas: data handling and general systems design. Among the improvements in data handling that will be necessary are random access mass storage devices of considerably greater capability than currently exists, better data relay capabilities, and "user friendly" data centers. These requirements are discussed in more detail below. Systems design efforts will have to focus on a variety of unmanned remote observatories and their instrumentation, as discussed above.

DATA MANAGEMENT AND ANALYSIS

Implementation of the scientific objectives and measurements of this strategy will give rise to a new set of data problems. Such problems will be the result of the generally increasing complexity of measurements and magnitude of data over the period for which the strategy is directed, and the global nature of the strategy. Adoption of this strategy will impose the significant requirement that the data chain, from observation to interpretation, must be well conceived and effective.

An overview of the measurement requirements specified in the discipline areas indicates a number of common data issues. It is apparent that over the next 20 to 30 years the increase in data volume will be due largely to meeting measurement requirements for high spatial and spectral resolutions, repetitive measurements, measurement of long-duration phenomena, and measurement of many short-duration phenomena with large-scale effects. The expected availability of advanced instrumentation, especially those that will provide more spectral bands and those that will produce large amounts of data, such as synthetic aperture radar, will add to the overall data volume.

As the science represented in these strategies matures, it is reasonable to expect that the complexity of measurement requirements and data interpretation will also mature. New sets and kinds of data will be requested, especially for the intercomparison of measurements of common sites or the comparison of widely separated sites. In this regard, the task group wishes to call particular attention to the area of data interpretation. The present ability and time required to translate space observations into analysis and understanding lag well behind the technology to build measurement devices and to collect data from space. To a large degree this lag is the result of overemphasis on an empirical approach to remote sensing. The effect has produced great delays not only in the interpretation and scientific understanding, but also in our ability to guide future measurements and instrument development based on the results of interpretation techniques. For this strategy, a major effort must be made in the development of new interpretation techniques in remote sensing for geology if the primary scientific objectives are to be met.

Beyond the need to reduce remote sensing data and to archive them in a common format is the need to apply a variety of analysis and image-processing techniques, to select among the data sets, and to combine the most appropriate data in a multispectral approach. An ongoing research effort should be maintained to develop optimum interpretational techniques for specific geologic problems. In oceanography, a similar research program in data interpretation techniques should be supported for new data acquisition systems. The development of such techniques must be accelerated to ensure that space measurements for earth science will produce the maximum scientific return within reasonable time periods following data acquisition.

The science objectives put forth here place no severe demands on U.S. capability to obtain data. There is no apparent technological barrier to the acquisition of data implicit in this strategy. The new precedent for data demands lies rather in the organization and management of data acquisition and analysis systems, and these new requirements are due in large measure to the now global extent of the earth sciences.

Driven by science requirements for ever more accurate and

more detailed remote sensing observations of the Earth, NASA has historically invested the lion's share of its resources in developing sensor, instrument, and platform technologies. Individual scientists have developed whatever methods they required and have acquired analysis tools from developers on an ad hoc basis. The scope and complexity of current and proposed remote sensing instruments, such as imaging spectrometers and synthetic aperture radar systems, promise to dramatically change the situation with respect to the science data analysis needs. Today's scientist must not only cope with a greater variety and complexity of data, but must also attempt to understand and utilize computer hardware and software technologies, which are themselves undergoing a phenomenal growth in capabilities. It is unreasonable to expect individual scientists to develop strategies and evaluate available resources for managing and analyzing all these data. If they have to do this, they will spend most of their time attempting to keep up with technology and have little time left to do science.

The earth science community feels that the time has come for NASA to play a significant role in providing the scientists with a program that will yield a coherent integration of data analysis and data management technologies. Although current support within NASA for data analysis/management technology is not insignificant, it is severely fragmented among a large number of different programs. In many cases the needs for data analysis are seen as only secondary to the goal of a particular program or project.

NASA currently has a strong ongoing program in space science data systems and in "pilot data systems" within its Information Systems Office. NASA also supports the NASA Ocean Data System (NODS) at JPL. As a result, many of the relevant issues such as data base management, storage, and communications technology are being addressed. The NASA EOS report recommends a further strengthening of these activities to include research, development, and implementation in the data analysis technology areas of (1) algorithm development and applied mathematics for analysis of large multispectral, multisensor, remote sensing image data and (2) utilization of the emerging technology of concurrent processing, with particular emphasis on high performance/cost ratio approaches, which hold the potential for putting large computational resources directly within reach of individual scientists.

An interagency study is one mechanism by which the requirements of a global data management for the next 25 years can be determined. This study should include at least the following elements: data processing, distribution, storage, and retrieval; data analysis needs; and computer capabilities. The task group suggests that this strategy for earth science and the CODMAC reports be utilized as a science requirement guideline for the study. In the interim, the task group urges the relevant agencies to begin to articulate the data issues and define the framework for establishing management and organization systems.

INTEGRATION OF RESULTS FROM OBSERVING SYSTEMS

The many experiments and instruments recommended in this chapter are difficult to relate directly to the major issues and themes set forth in Chapter 3. In this section the task group briefly reviews these leading themes and suggests how the measurement programs would contribute to improved insight. In a few cases, the programs will generate specific diagnostics for certain questions. But in most instances their contribution is to give a richer context to the ongoing scientific thinking about the Earth, which stimulates important new ideas—the most important often being the most unpredictable.

Studies of the solid Earth will be advanced most significantly by the geodetic and seismological systems proposed in this chapter. Detailed complexes of geodetic measurements and seismometers will give a mapping of crustal motions in zones of deformation, such as the San Andreas, necessary to a greatly improved insight into the occurrence of earthquakes and other processes that determine the structure of the lithosphere. Practical by-products of their insight will be effective earthquake prediction and clarification of the tectonic contexts of mineral formation. On a more globally interconnected scale, seismological networks with satellite gravimetry will produce more detailed maps of density variations throughout the mantle. These maps, coupled with computer models, will determine the nature of mantle convection, the spectrum or length scales of its flows, the degree of interaction among different parts of the mantle, and so forth.

The programs of space measurements and associated observations will give a more detailed description of a wide range of phenomena at both regional and global scales. The descriptions will enable greatly improved short-term predictions and insights into ongoing processes. In several respects, the models for these processes will be largely separate. In particular, this separation exists between solid earth processes on the one hand, and oceanic and atmospheric processes on the other. The coupling of these different subsystems occurs on a much longer time scale: millions of years. The nature of the hydrosphere and atmosphere depends on the outgassing of the mantle. Hence, a more fundamental problem is the long-term evolution of the mantle, which will depend strongly on convection that in itself was, going back in time, increasingly different from the contemporary convection whose manifestations are measured by current geodetic and seismological systems.

The global data sets to be collected through EOS and PLATO, together with modeling on high-speed supercomputers, will allow for the first time the integration of processes on many time scales and from different disciplines. Only in this way can we address the wide disparity of time and space scales represented by geophysics, geochemistry, fluid dynamics, and biological processes on Earth.

6
Science Policy Considerations and Recommendations

INTRODUCTION

In order to make progress on the grand themes, the strategic approach described in this report must be implemented with a specific program that encompasses global, long-term satellite measurements, complementary in situ observations, and concurrent modeling on state-of-the-art computers. The type of system the task group envisions will provide data at a far greater rate than at present; major efforts in technology and coordination will be required to adequately process and distribute the data to the research and operational communities. For this system to operate efficiently and to be cost-effective, there must be coordination and cooperation both on the national level among those agencies involved with earth science and at the international level among all those countries that participate in either satellite or in situ programs.

RECOMMENDATIONS

The task group is cognizant of the many significant policy issues associated with carrying out all of the elements of a Mission to Planet Earth. Although a complete discussion of these issues is beyond the scope of this report, there are several important policy areas that merit comment.

1. National Coordination of Agency Roles

The task group recommends that full coordination of the U.S. federal agencies involved in the civil earth science effort be developed at both the programmatic and the budgetary level. Full cooperation among the agencies involved in the civilian earth science effort—NASA, NOAA, NSF, USGS, DOD, and others—is required to develop effective, coordinated programs and budgets in proportion to assigned tasks. The assignment of the appropriate relative roles of NASA as a research and development agency, NSF as a supporter of basic research, and NOAA and USGS as operational, mission-oriented agencies in earth sciences is a test case. If significant responsibilities are shifted to NOAA, then new funding will be required there as well as assiduous policy attention. The task group recognizes the importance of establishing clear roles, especially as we look to measurements on longer and longer time scales. Moreover, coordination with the commercial sector is essential, and any comprehensive program must include this sector as a major player.

The absence of a well-coordinated organizational structure at the federal level for conducting global earth science research has several symptoms:

(a) The distinction between "research" and "operations" or "monitoring" is sometimes artificially forced, which has negative consequences. For instance, data taken by monitoring services that would be valuable for research purposes are not retained, or are retained in aggregated form, thereby obliterating possibly significant details; improved insights and procedures resulting from research do not affect operations as much as they should; and long-term measurements necessary for research may not be continued. These problems tend to arise most strongly when there is a division of responsibility among organizations. In the earth sciences, this is most notable between NASA and NOAA, but it also arises in activities involving DOD, NSF, and USGS.

Particular attention should be directed to matters that may fall through the cracks because of the distinction between "research" and "operations," or other distinctions generated by non-scientific considerations. Examples include climatological studies that use regularly taken surface data; surface observation systems that require data transmission facilitated by spacecraft; data taken by DOD satellites that are currently unavailable but may become available later with a change of policy (e.g., radar altimetry by Geosat); measurements on the Earth's surface necessary to make effective use of space missions for certain purposes (e.g., geological and ecological "ground truths" needed to utilize Landsat for these purposes); and observations from the Earth's surface using spacecraft that are no longer, or never were, active NASA projects, such as tracking of LAGEOS and NAVSTAR to obtain geodetic positions for crustal motion studies.

(b) Traditionally, different levels of funding among agencies have often led to severe imbalances in space-related programs and thence to pressure for NASA to undertake aspects of the work that might normally be the province of another agency. An example is the stratospheric ozone program, in which NASA's level of activity relative to other agencies is much higher than contemplated in the charge by Congress. There appears to be a need for more thorough oversight by the Office of Science and Technology Policy (OSTP) in concert with OMB budget review to assure adequate funding in other agencies for a balanced effort.

(c) Interagency cooperation is essential to the advancement of the earth sciences, yet such cooperation in the area of satellites has not fared well at OMB. For example, the NSCAT program logically requires a NOAA contribution for civilian data distribution, and GRM would benefit from USGS interaction. Yet the prevailing mood of OMB has been that if an agency sees a major role in a particular satellite program, that agency should pay the full cost. Satellites are expensive and cannot be absorbed into existing agency budgets—new funds must be found. For each satellite program required to make a major new impact in our understanding of the Earth, the appropriate agencies should receive a fair hearing as to their need for new funds. A joint budget review for those earth-science-related agencies (NOAA, USGS, Navy) under cabinet departments, together with independent agencies (NASA, NSF, EPA) may be the best solution.

(d) There is great variability within NASA in utilization of the university and industrial research communities, and consequently in the quality of research and in the multiplicity of approaches important to assure a reasonably optimum solution. The earth sciences in NASA lag significantly in these matters behind planetary sciences, astronomy, and space physics, perhaps because of the "applications" tradition. Progress is being made in developing peer-reviewed grant programs in several areas, but it is uneven. Some activities have been sheltered from this competition on the argument that they are unique functions necessarily done in-house. While it is correct that some research and technology must be done in-house, the more "unique" such activity is, the greater is its need for close scrutiny by the best experts available to assure that it is done well.

(e) It is the task group's understanding that a reexamination of agency roles in earth-oriented space science is currently under way in the federal government, and that an enlarged scientific role is being explored for NOAA, in addition to its operational responsibilities. For such a shift to be effective, NOAA must develop a much more intensive interaction with the relevant scientific community in universities and elsewhere. This intense interaction in turn requires much more support and encouragement from the administration than has been evident in funding and personnel decisions affecting NOAA since 1976.

2. International Coordination and Cooperation

The task group recommends that steps needed to assure the required level of international participation and cooperation be taken at the early stages of program planning.

Earth science programs in other countries and international agencies are strong, and expected to become stronger. The coordination of the many international earth observing spacecraft scheduled to fly before the year 2000 is one of the most important and beneficial objectives for the earth sciences. The United States must view its own program as one contribution to an overall international program—both to obtain access to active scientific and technical communities, and to develop the global system of in situ observations. International participation is needed to support space observing systems and to deploy and operate in situ measuring devices. Site selection, standardization of equipment, data rates and formats, and data exchange protocols are examples of the issues that must be resolved to ensure the success of Mission to Planet Earth. There is an excellent record of international cooperation in large-scale projects; examples include the International Geophysical Year of 1957, the Geodynamics Program, the Global Atmospheric Research Program, the International Lithosphere Project, the Ocean Drilling Programs, the World Climate Research Program, and the Man and the Biosphere Program. The International Geosphere-Biosphere Program, which builds on the developing concepts of global habitability and biogeochemical cycles and is aimed at an understanding of global change, is still in the planning stage. The program envisioned here is designed to provide the long-term global synoptic data required to understand the whole earth system, including all of the disciplines in the programs mentioned above.

3. NASA's Role in the Solid Earth Sciences

NASA is to be commended for the strong role it has played to date in earth sciences, ranging from atmospheric to oceanic to land surface processes to solid earth sciences. The use of satellites in studies of the Earth's interior is as old as the space program. From the beginning, satellites were used to more precisely define the figure of the Earth and to investigate the distribution of mass in its interior. The intervening decades have seen the development of NASA satellite missions and ground systems that have been more and more specifically designed for planetary-scale, solid earth science measurements. The technological advances on which these measurements have been based were the products of, or directly linked to, space technology. NASA should continue to play this key role in the development of space-based technology and the complementary earth-based systems, together with the necessary data systems, as new observatories for the Earth are developed.

NASA has been able to broaden the base of its solid earth science studies through interagency agreements. At the same time, *the task group recommends that NASA develop and support a more comprehensive program in the solid earth sciences within NASA itself.* In developing such a program, NASA should not constrain itself to the use of space technology, but must be prepared to accept direct programmatic and financial responsibilities covering a broader spectrum of research activities in solid earth research. The task group endorses the position of the Earth Observing System (EOS) Science and Mission Working Group, which recognized that in future NASA missions, "satellite-obtained data must be used in concert with data from more conventional techniques," and noted that, in addressing multidisciplinary problems, "observational capabilities must be employed which range in scale from detailed in situ and laboratory measurements to the global perspective offered by satellite remote sensing."

This philosophy fits perfectly with NASA's role in planetary sciences and indicates the major role that NASA can take in studies of the Earth as a planet. While NASA has developed a comprehensive program in the planetary sciences, such plans have never included the Earth, in spite of the fact that it is impossible to design a program of real value for solid earth science that does not involve all of the requisite studies. We have an opportunity to define a comprehensive program that deals with all of the most exciting and most important questions in the solid earth sciences today: questions such as the primary differentiation of the Earth, the origin of magmas, the driving forces for plate tectonics, and the generation of the Earth's atmosphere. NASA's engineering capability in advanced technology, such as satellite systems and data base management, is essential to the accomplishment of these objectives. Clearly, such studies must be carried out with the other agencies that support basic research in earth sciences—notably NSF, USGS, and NOAA—but a strong program in NASA itself must be maintained.

Appendix A
EOS Instrument Descriptions

MODIS-T MODERATE RESOLUTION IMAGING
          SPECTROMETER—TILT
          Imaging spectrometer for the measurement of biological and physical processes on a 1 km × 1 km scale. Scanning instrument covering a 1500-km swath centered at nadir, with a 0° to 30° scan mirror tilt for sun avoidance. Spectral range of 0.4 to 1.04 μm in 64 bands, any 17 selectable for telemetering.

MODIS-N MODERATE RESOLUTION IMAGING
          SPECTROMETER—NADIR
          Imaging spectrometer for the measurement of biological and physical processes on a 1 km × 1 km and a 0.5 km × 0.5 km scale. Scanning instrument covering 1500-km swath centered at nadir. Spectral ranges of 0.4 to 2.3 μm, 3 to 5 μm, and 6.7 to 14.2 μm. Atmospheric sounding channels under study. Total of 40 bands, 27 at 1-km and 13 at 0.5-km resolution.

PRECEDING PAGE BLANK NOT FILMED

Descriptions courtesy of NASA.

HIRIS   **HIGH RESOLUTION IMAGING SPECTROMETER**
Spectrometer providing highly programmable localized measurements of biological and geological processes at a spatial resolution of 30 m. Pointable instrument with a 50-km-swath coverage, viewing accessible
areas up to 30° off track. Spectral coverage of 0.4- to
2.5-μm at 10-nm resolution. 210 spectral bands.

TIMS   **THERMAL IMAGING SPECTROMETER**
Spectrometer providing high-resolution thermal imaging capability to the HIRIS for biological and geological measurements. Instrument will be bore-sighted
with HIRIS to provide 30-m resolution over a 50-km
swath. Its spectral range will cover from 3.0 to 5.0 μm
and from 8.0 to 14.0 μm, for a total of 8 bands of 50-
and 500-nm spectral resolution.

AMSR   **ADVANCED MECHANICALLY SCANNED
RADIOMETER**
Radiometer providing measurements of atmospheric
water vapor, precipitation, and snow and ice extent.
Rotating antenna assembly scans the Earth in the forward velocity direction ±75° to the subsatellite track
for a swath of 1400 km. Resolution varies from 20 km
to 2 km over the six frequencies of 6.0, 10.65, 18.7,
21.3, 36.5, and 91 GHz. Both H and V polarizations
are available. Antenna size is approximately 4 m × 7
m.

ESTAR   **ELECTRONICALLY SCANNED THINNED
ARRAY RADIOMETER**
Microwave radiometer providing high-resolution soil
moisture measurements using aperture synthesis techniques. Single-frequency instrument at 1.4 GHz. Antenna size 18 m × 18 m covering a swath of 900 km.
Approximate temperature resolution of 1K.

AMSU   **ADVANCED MICROWAVE SOUNDING UNIT
(A and B)**
Microwave radiometer providing measurements of atmospheric temperature and humidity. Twenty-channel
instrument divided into AMSU-A and AMSU-B subsystems. AMSU-A primarily provides atmospheric

temperature measurements from the surface up to 40 km in 15 channels; i.e., channels at 23.8 GHz, 31.4 GHz, 12 channels between 50 and 60 GHz, and 89 GHz. Coverage is approximately 50° on both sides of the suborbital track, with an IFOV of 50 km. Temperature resolution equivalent to 0.25 to 1.3K. AMSU-B primarily provides atmospheric humidity measurements in 5 channels at 89 GHz, 166 GHz, and 183 GHz (3 channels). Coverage is the same as AMSU-A, with an IFOV of 15 km and a temperature resolution of 1.0 to 1.2K.

LASA   LIDAR ATMOSPHERIC SOUNDER AND ALTIMETER
Lidar system(s) for the measurement of total column content water vapor, water vapor profiles, aerosols and clouds, atmospheric state parameters, ozone, ice sheet mass balance, and land topography. System initially uses a 1.25-m-diameter telescope and multiple-wavelength selective laser systems tailored to specific measurement areas. Telescope is a fixed nadir viewing system, which may be modified for scanning and/or aperture size change. Growth version of the SISP LASA for the measurement of tropospheric constituents and atmospheric-state parameters. Multiple-wavelength meter class telescope/laser system with pointing and/or scanning capability. Water vapor and ozone profiles together with pressure and temperature measurements are system goals.

SAR   SYNTHETIC APERTURE RADAR
Imaging radar for all-weather studies of surface processes for land, vegetation, ice, and oceans. Resolution of 30 m over a swath width of approximately 100 m in width from 20° to 60° on either side of the subsatellite track. C, L, and X band frequencies with H and V polarizations.

ALT   ALTIMETER
Microwave system for the study of ocean topography

and topographical measurements over ice. Two-m-diameter antenna for 10-cm precision height measurement over the oceans. System operates at 13.5 GHz with a subsatellite swath of 2 to 15 km. N-ROSS capability of less than 5-cm altitude accuracy, 0.5-m or 10 percent significant wave height precision, 2-m/s wind speed accuracy.

SCATT   SCATTEROMETER
Microwave system for the measurement of wind stress over the oceans. System provides measurement along two parallel swaths from 120 km to 700 km on either side of the subsatellite track. System has a 6-stick antenna configuration, 1-m/s wind speed and 10° angular resolution are goal measurements.

CR   CORRELATION RADIOMETER
Instrument for measurement of constituents in the troposphere using selective absorption gas cell radiometry. Nominal 4.3° field-of-view, ±5° to nadir. Provides global distribution of carbon monoxide and other species, with limited vertical resolution.

NCIS   NADIR CLIMATE INTERFEROMETER SPECTROMETER
Michelson interferometer-type instrument for the measurement of temperature profiles and tropospheric constituents, i.e., H_2O, NO_3, CH_4, etc. Approximate spectral coverage of 6.0 to 40 μm. Nominal 5° field-of-view.

LAWS   LASER ATMOSPHERIC WIND SOUNDER
Lidar system for direct tropospheric wind measurements. System consists of a 1.25-m-diameter rotating telescope (3 rpm) with a half-cone angle of 53°. Instrument covers a swath area of 300 km × 300 km for wind measurements of 1 m/s.

IR-RAD   INFRARED RADIOMETER
Radiometer for the measurement of upper atmospheric temperature, winds, and certain atmospheric species. A multiband, cryogenically cooled instrument, which views the limb over an approximate 100° × 1.9° scan

field-of-view. Measures temperature, ozone, and trace species distribution with approximately 3-km vertical resolution.

PMR   PRESSURE MODULATED RADIOMETER
Radiometer measuring emissions from selected atmospheric constituents and providing global maps of atmospheric temperature and constituents abundance. Multichannel infrared instrument using pressure-modulated gas cells for detection of selected species. Views the Earth's limb perpendicular to the flight path. Approximate resolution is 2.6 km at the limb.

MLS   MICROWAVE LIMB SOUNDER
Microwave radiometer that detects and measures the distribution of trace chemical species in the upper atmosphere. Radiometer is a multifrequency antenna system, which scans the Earth's limb and detects the thermal emission of trace species at selected frequencies of 83, 119, 183, 205, and 231 GHz. Nominal vertical FOV of 3 to 10 km; horizontal FOV of 10 to 30 km at the limb.

SUB-MM   SUBMILLIMETER SPECTROMETER
Advanced spectrometer technology that detects and measures the distribution of trace chemical species in the upper atmosphere. Instrument is a scanning heterodyne system that views the limb with an FOV of approximately 1.8° to 80°. Improved resolution over the MLS. Measures trace species such as OH and HCL.

VIS-UV   VISIBLE-ULTRAVIOLET SPECTROMETER
Multiple spectrometer system for measuring trace species in the upper atmosphere using selected emission lines. Instrument is a multiple-grating, high-resolution (3 to 10 Å) spectrometer covering the wavelength range of from 300 to 12,000 Å.

F/P-INT   FABRY-PEROT INTERFEROMETER
Interferometer system measuring wind and temperature in the upper atmosphere. Instrument consists of a multiple-etalon interferometer fed by a gimbaled telescope, which views the limb in the 10- to 50-km

118

altitude region to measure the absorption features of
oxygen and scattered light bands and to measure the
atmospheric emission features in the 60- to 300-km
altitude region. Measures the vector wind field to less
than 5 m/s.

CIS **CRYOGENIC INTERFEROMETER
SPECTROMETER**
A combined cryogenically cooled Michelson interferometer and multichannel radiometer (Ebert type),
sharing a common narrow field-of-view telescope. Instrument views the Earth's limb and measures emissions from trace constituents and temperature in approximately five wavelength bands.

SCM **SOLAR CONSTANT MONITOR**
Instrument for the precise solar total irradiance monitoring during the solar cycle. Active cavity radiometer-
type instrument. Accuracy of 99.9 percent at solar
total irradiance level. Wavelength coverage of 0.001 to
1,000 μm. Total irradiance of 0 to 200 W/m^2. Field-
of-view of $\pm5°$ with $\pm1°$ for solar pointing.

SUSIM **SOLAR UV SPECTRAL IRRADIANCE MONITOR**
Instrument system for the measurement of solar UV
fluxes and their changes over a solar cycle. Full disk
solar spectral irradiance measurement. Two double
dispersion scanning spectrometers, one for measuring
solar flux and the second to measure the output of
calibration lamps. Wavelength coverage from 120 to
400 nm with spectral resolution of 0.1, 1.0, and 5.0
nm.

ERBI **EARTH RADIATION BUDGET INSTRUMENT**
Multiple-channel instrument for the measurement of
the Earth's radiation budget on a regional, zonal, and
global scale, using selected fields of view. Instrument
response is from 0.2 μm to 50 μm with selective filters.
Instrument uses the Sun as a calibration source. The
instrument is divided into a scanning and nonscanning
portion for the selective measurements.

AVHRR/2 ADVANCED VERY HIGH RESOLUTION
RADIOMETER (MOD 2)
Visible/infrared radiometer for the measurement of
cloud cover, sea surface temperature, and the mapping of ice and snow. Also used to measure vegetation cover. Five-band radiometer system covering the
spectral range of 0.58 to 12.5 μm, i.e., 0.58 to 0.68,
0.7 to 1.1, 3.55 to 3.93, 10.3 to 11.3, and 11.5 to 12.4
μm. 1-km IFOV, 1600-km swath, centered at nadir.
Sea surface temperature resolution of less than 1K.
AVHRR/3 operable on NOAA-K through M will have
a time-shared modified channel 3, i.e., 3a 1.57 to 1.78,
3b, 3.55 to 3.93 μm for cloud cover detection. Also
channel 2 will change to 0.82 to 0.92 μm to eliminate
a water vapor absorption.

MRIR   MEDIUM RESOLUTION IMAGING RADIO-
METER
Ten-band scanning radiometer for improved sea surface temperature, cloud mapping and the monitoring
of surface processes in vegetation coverage. Four additional bands to the AVHRR/3, i.e., 0.45 to 0.52, 0.52
to 0.60, 2.08 to 2.35, and 6.5 to 6.7 μm. Improved spatial resolution of 250 to 500 m, with a ground swath
of 2230 km.

ATSR   ALONG TRACK SCANNING RADIOMETER
IR radiometer with a scanning technique that views
the same point from two different angles, providing an
improved atmospheric correction. Three-band system
at 3.7, 11, and 12 μm. 1-km spatial resolution with
a 500-km-swath coverage. Temperature resolution approximately 0.1K.

ARGOS   FRENCH DATA COLLECTION and LOCATION
SYSTEM
(ADCLS) Data collection and location system for drifting platforms and balloons. 401.650-MHz operating frequency
for a drifting buoy location accuracy of 1 km and speed
determination of 0.3 m/s.

GOMR — GLOBAL OZONE MONITORING RADIOMETER
An atmospheric sounding instrument measuring radiances and solar IR radiances at the top of the atmosphere to provide ozone column amounts, ozone vertical profiles, and solar UV fluxes. Instrument operates in 13 bands over the wavelength region of 273.5 to 380 nm. The GOMR has a nadir FOV of 11.3°, equivalent to a spatial resolution of 169 km.

HIRS/2 — HIGH RESOLUTION IR SOUNDER (MOD 2)
Scanning radiometer for temperature and moisture profiles. Twenty-band system covering the wavelength regions of 3.76 to 14.96 μm with an albedo measuring channel centered at 0.69 μm. Cross-track scan is ±49.5°, equivalent to a ground swath of 2230 km centered at nadir. Ground IFOV is 20 km in diameter. HIRS/3 mod will change the band pass of channel 20 to a broad response, i.e., 0.3 to 1.1 μm or 0.4 to 1.8 μm.

MLA — MULTILINEAR ARRAY
Conceptual system consisting of targetable and global coverage instruments for the study of land processes.
Global Repetitive Coverage Seven-band instrument covering the wavelength regions of 0.45 to 0 9, 1.55 to 2.35, and 10.4 to 12.5 μm. Thirty-meter resolution in the visible, near, and middle infrared regions. Sixty-meter resolution in the thermal infrared. 185-km swath width.
Targetable Coverage Thirty-two-band (8 selectable) instrument covering the wavelength regions of 0.45 to 1.1 and 1.2 to 2.4 μm. Ten-meter resolution in the visible and near infrared, 20-m resolution in the middle infrared. Stereo and cross-track viewing ±30° nominal.

SEM — SPACE ENVIRONMENTAL MONITOR
(ADVANCED CONFIGURATION)
A group of seven sensors for in situ measurement of plasma parameters.
Total Energy Detector (TED) Measures total energy deposited by magnetospheric electrons and protons in the range of 0.3 to 20 keV.

Medium Energy Proton and Electron Detector (MEPED) Four directional sensors and an omni-directional sensor for measuring protons, electrons, and ions in the range of 30 to greater than 60 keV.

Precipitating Proton-Electron and Cumulative-Dose Spectrometer Measures electron dose over the 1- to 10-MeV range and protons of greater than 20 MeV. The spectrometer measures the precipitating electrons and ions in 20 energy bands between 30 eV and 30 keV.

Ionospheric Plasma Monitor Measures ambient electron and ion density and temperature.

Scanning Gamma- and X-Ray Sensor Measures x-ray intensity as a function of energy from 2 keV to 100 keV.

X-Ray Energy Detector Measures x-ray energies in the range of 25, 45, 75, and 115 keV.

High Frequency Receiver Monitors the ionospheric noise breakthrough frequency used in ionospheric forecasting.

GLRS   GEODYNAMICS LASER RANGING SYSTEM
Laser Ranging System for the study of crustal movements. Range measurements made in tectonically active regions and across tectonic plate boundaries using arrays of retroreflector targets acquired by the satellite system. System uses frequency double-mode locked Nd:YAG laser: 20 millijoules at 532 nm, 0.1 milliradians divergence with a 10 pulses/second repetition rate. Receiver telescope has a diameter of 18 cm and a ranging/tracking precision of 0.01 milliradians. Ground spot diameter, 80 m at nadir. Data is used to monitor intersite distances associated with crustal motion.

www.ingramcontent.com/pod-product-compliance
Lightning Source LLC
Chambersburg PA
CBHW051320170526
45166CB00002B/615